Q Fever

RIVER PUBLISHERS SERIES IN RESEARCH AND BUSINESS CHRONICLES: BIOTECHNOLOGY AND MEDICINE

Volume 8

Indexing: All books published in this series are submitted to Thomson Reuters Book Citation Index (BkCI), CrossRef and to Google Scholar.

Combining a deep and focused exploration of areas of basic and applied science with their fundamental business issues, the series highlights societal benefits, technical and business hurdles, and economic potentials of emerging and new technologies. In combination, the volumes relevant to a particular focus topic cluster analyses of key aspects of each of the elements of the corresponding value chain.

Aiming primarily at providing detailed snapshots of critical issues in biotechnology and medicine that are reaching a tipping point in financial investment or industrial deployment, the scope of the series encompasses various specialty areas including pharmaceutical sciences and healthcare, industrial biotechnology, and biomaterials. Areas of primary interest comprise immunology, virology, microbiology, molecular biology, stem cells, hematopoiesis, oncology, regenerative medicine, biologics, polymer science, formulation and drug delivery, renewable chemicals, manufacturing, and biorefineries.

Each volume presents comprehensive review and opinion articles covering all fundamental aspect of the focus topic. The editors/authors of each volume are experts in their respective fields and publications are peer-reviewed.

For a list of other books in this series, visit www.riverpublishers.com

Q Fever

Svetoslav P. Martinov

National Diagnostic and Research Veterinary Medical Institute – Sofia
Bulgaria

LONDON AND NEW YORK

Published 2017 by River Publishers
River Publishers
Alsbjergvej 10, 9260 Gistrup, Denmark
www.riverpublishers.com

Distributed exclusively by Routledge
4 Park Square, Milton Park, Abingdon, Oxon OX14 4RN
605 Third Avenue, New York, NY 10158

First published in paperback 2024

Q Fever / by Svetoslav P. Martinov.

Routledge is an imprint of the Taylor & Francis Group, an informa business

Publisher's Note
The publisher has gone to great lengths to ensure the quality of this reprint but points out that some imperfections in the original copies may be apparent.

While every effort is made to provide dependable information, the publisher, authors, and editors cannot be held responsible for any errors or omissions.

ISBN: 978-87-93519-49-7 (hbk)
ISBN: 978-87-7004-440-0 (pbk)
ISBN: 978-1-003-33915-1 (ebk)

DOI: 10.1201/9781003339151

Contents

Preface

Q fever is a zoonosis caused by *Coxiella burnetii*, an obligate intracellular micro-organism. The disease has a worldwide distribution. The natural reservoirs for this agent are a number of domestic and wild animal species. Aerosols play the most important role in the transmission of the infection to humans. A number of characteristics and properties of the pathogen and the disease, define Q fever as a difficult and lasting veterinary and medical problem. *Coxiella burnetii* is currently ranked as a "Category B bioterrorism agent".

The aim of this book is to describe the biological, morphological and immunological properties of *C. burnetii*, the clinical forms of the disease, the pathogenesis and pathology, the epidemiological peculiarities in animals and humans, sensitivity to antibiotics and treatment ant the prevention and control.

The unique nature of the causative agent is analyzed from the standpoint of the contemporary microbiology, cell biology and molecular biology. The host range and spread of the *C. burnetii* infections in mammals, birds, ticks and man are reviewed and detailed. The epizootiological and epidemiological particularities and relationships are emphasized. Special attention has been paid to both types of foci of the infection – agricultural and natural.

The accepted and proven methods for the isolation, cultivation, identification and differentiation of *C. burnetii* from materials with different origins have been presented. The book shows well documented data for the direct demonstration of the pathogen by means of electron microscopy and the histopathology in domestic ruminants and certain laboratory animals affected by Q fever.

The etiologically proven clinical forms of Q fever have been divided in seven groups. The sensitivity of strains *C. burnetii* to certain antibiotics depending on its phase condition have also been explored.

Finally, the issues related to the prevention and control of Q fever have been assessed and a system for monitoring, prophylaxis and fight against this zoonotic disease has been proposed.

The book emphasizes that the efficient control of Q fever zoonosis requires a close cooperation between the veterinary and medical authorities.

Numerous references had been selected for providing original information and further sources to details of the issues discussed.

I have intended the book for the specialized audience of veterinary and medical professionals and students, however, due to the worldwide implications of the disease, the book would be also very informative to a broader range of readers.

Professor Svetoslav P. Martinov, DVM, Ph.D., D.Sci.

List of Figures

List of Tables

List of Abbreviations

AE	Antigenic unit
AF	Agricultural foci
BHK-21	Baby hamster kidney-21
Bp	Base pairs
CC	Cell cultures
CE	Chicken embryos
CEF	Chicken embryonic fibroblasts
CF	Complement fixing
CFT	Complement fixation test
DEM	Direct electron microscopy
DNA	Deoxyribonucleic acid
EM	Electron Microscopy
GMT	Geometric mean titer
IFA	Indirectimmunofluorescence
IFHT	Immunofluorescence hemocytic test
i.m.	intramuscular inoculation
i.p.	intraperitoneal inoculation
i.v.	intravenous inoculation
LCV	Large cell variant
LM	Light microscopy
MIC	Mean inhibitory concentration
MIFT	Microimmunofluorescence test
MST	Mean survival time
NBM	New born mice
NF	Natural foci
nm	nanometer
OIE	World Organization for Animal Health
PCR	Polymerase chain reaction
PHYLYP	Phylogeny inference package
RAPD	Randomly amplified polymorphic DNA analysis

REP	Repetitive extragenic palindromic sequence polymorphism
s.c	subcutaneous inoculation
SCV	Small cell variant
UPGMA	Unweighted pair group method
YS	Yolk sac

Introduction

Q fever is a zoonotic disease with global distribution and important health, social, and economic significance. It is caused by highly infectious pathogen Coxiella burnetii. A number of other properties and characteristics of the causative agent and disease make Q fever as a lasting and difficult veterinary and epidemiological problem, namely the adaptability of C. burnetii and its high resistance in the external environment; the possibility of the existence of the agent in three- and two-member parasitic systems; the availability of natural and agricultural foci of infection; peculiarities of pathogenesis in humans and animals, and the mechanisms of excretion of the pathogen into the environment; and the high susceptibility of non-immune populations of animals and people. Given that C. burnetii is included in the arsenal of bacteriological weapons as an agent with a potential bioterrorist threat, it must be borne in mind the strategic importance of this microorganism. Although the infectious agent and Q fever were discovered long ago, they are still insufficiently studied and rickettsiology is little known to the wider circles of microbiologists, virologists, and clinicians. For example, there is no complete understanding yet of the various clinical forms, which manifests the Coxiella infection in animals and humans, and which can lead to undetectable cases, incomplete clinical-epidemiological analysis and conclusions, and difficulties in the treatment and the reduction of the disease. The perceptions about the biology of C. burnetii and the pathogenesis of Q fever are still hypothetical and need a continued most in-depth research, including the level of electron microscopy and molecular biology. Etiological diagnosis is a key issue, which is the subject of numerous studies. The isolation and cultivation of the pathogen is an important scientific and practical problem with specifics regarding the sensitivity of the biological models, the duration of laboratory procedures for proof of the cause and for the safety of the personnel. There was a need to adopt new approaches to the classic methodology for culturing C. burnetii, a proper treatment of the inocula to ensure a high multiplicity of the infection, and the introduction of methods for the early indication and identification of isolated strains. Particular attention is paid to the specifics of the cultivation of C. burnetii in cell cultures due to the difficulty of the penetration of the agent into the cell. The morphological proof of this microorganism is based on the traditional method of light microscopy, accompanied by serious shortcomings and difficulties. Electron microscopy is widely used for the diagnosis of viruses and chlamydiae, while in coxiella and rickettsia, it has not been developed. Important in the studies of C. burnetii and coxiellosis are the electron microscopy for the early indication and identification of isolates in chicken embryos, cell cultures and laboratory animals, and direct electron microscopy of the pathogen in clinical, pathological, and experimental materials from mammals and birds. These investigations provide opportunities for studying the ultrastructure and the reproductive cycle of C. burnetii and the pathological changes in infected cells. They can throw light on the tropism of the agent in the host and disease pathogenesis. The modern molecular biological methods to study C. burnetii have also been used in laboratory practice, mainly for research purposes, with the prospect of a wider use in diagnostics and molecular epidemiology.

Sheep, goats, and cattle are very susceptible to infection with C. burnetii. They show a certain sensitivity and other domestic animals as their susceptibility to infection and its scope is seen as a dynamic characterization requiring systemic supervision. It is imperative for a full disclosure of the nosological geography of Q fever, which is one of the leading in importance and presence zoonosis in different countries. Important mechanism in this regard is to have the complex serological and virological tests for the detection of new outbreaks of infection and for the tracking of the dynamics of previous ones throughout the country. The detection of infected animals should not be only on species base, but also on clinical indicators. The concept of latent and inapparent dominant character of Q fever in animals enduresa serious correction based on accumulating evidence of clinically manifested disease, often having a mass character. Coxiella-induced abortions, stillbirths, and non-viable offspring in domestic ruminants occupy a growing share in the pathology of these species. The same applies for Coxiella-induced respiratory diseases, mastitis, endometritis, infertility, and syndromes accompanied by prolonged reduction of milk production. C. burnetii and its clinical manifestations are of great interest also in other types of domestic mammals, including rabbits, dogs, cats, and poultry. Obviously, the clinical aspect of animal coxiellosis needs thorough scientific elucidation on a broad experimental basis. In close connection with this question is the directed drug therapy based on the use of tetracycline antibiotics. Despite the relatively good results, the scientific research is still aimed at investigating more effective antibiotic and chemotherapeutic agents.

The remaining unsolved problem is the lack of reliable and effective vaccines for the immunization of the disease in animals.

The key moment in the epidemiology of Q fever is the existence of two types of foci of infection with C. burnetii—natural and agricultural. Both types of foci are dynamic multicomponent systems with a number of features and relationships, the condition of which requires constant supervision and scientific approach. Since 1989, in Bulgaria and other countries radical reforms have been carried out affecting the ownership structure and the agricultural economics—processes that inevitably lead to changes in the epidemiology of the infectious diseases and come accompanied by the emergence of new risks. This new situation requires adequate science-based answers and a rethinking of the current approaches to monitoring, prevention, and control. An in-depth study of the current state paired with a historical comparison of the agricultural reservoir of Q fever, the determination of the proportion of the different species in its formation, the improvement and refinement of the territorial map of the spread of the disease among livestock, and the overall analysis of the epidemiological features are urgent tasks of fundamental and applied nature.

Another major trend in the modern epidemiological studies of Q fever is natural foci of infection that requires an updated assessment of the state of reservoirs and vectors of C. burnetii in nature—ticks, wild mammals and birds, in terms of species coverage, extent of infection, activity, territorial distribution, and the available etiological and epidemiological connections inside the foci. In parallel, there should be performed essential studies on the relationships and interdependencies between the agricultural and natural foci, on new data for single etiology of the diseases caused by C. burnetii in animals and humans, and on the relationship between the intensity of the outbreaks of animal diseases and the epidemics in humans.

The information on the changing epidemiology and expanding no so logical range of Q fever in humans deserves a strong attention.

The analysis of the results of research on C. burnetii and Q fever shows that the pathogenis still insufficiently studied, and the infection becomes more wide spread among mammals, birds, and people.

The timeliness and many obscure aspects of the problem, of course, necessitated the expansion and the development of the advanced level of new issues relating to agent and disease in different susceptible species. They are reflected in the statement of the book summarizing the scientific facts and developments internationally, contributions of Bulgarian researchers and longtime personal experience of the author. Special attention is paid to the state of Q fever in Bulgaria and advanced control capabilities, prevention, and control.

I believe that this monograph "Q fever" will enrich the knowledge of the disease of the medical and veterinary professionals, microbiologists, epidemiologists, molecular biologists, and students in human and veterinary medicine. This work will provide current information on zoonotic problem with medical, social, and economic significance and a wider audience of society as a whole.

1

Historical Notes

The first clinical description of Q fever in humans was done by Edward Derrick in Australia in 1937 [164]. The author reported febrile illness of unknown etiology among abattoir workers in Brisbane, Queensland state. The name of the disease comes from the first letter of the word querry (Eng)— "Vague, indefinite," which underlines its ambiguous nature (Querry fever; Q-fever). Derrick inoculated guinea pigs with blood and urine of sick people and caused fever, but was unable to determine the pathogen, which makes the assumption that it was a virus. Supplied by Derrick material (emulsion of guinea pig spleens), other Australian researchers—Burnet and Freeman [117]—reproduced the disease in guinea pigs and observed in the spleen intracellular vacuoles filled with granular formations (hematoxylin–eosin staining) and many small sticks (Castaneda or Giemsa staining). In 1939, the causative agent of Q fever is named Rickettsia burnetii. In the same period, the American scientist Gordon Davis performs independent research and found that ticks of the species Dermacentor andersoni collected in Nine Mile, Montana state, and put on guinea pigs cause febrile illness other than the disease Rocky Mountain spotted fever. Later study joins Herald Cox, which further characterized the microorganism ("Nine Mile agent") as both similar to viruses and rickettsiae and able to pass through filters [149, 155], for which he was named R. diaporica. Soon afterward, Cox and Bell [150] cultivated the pathogen in chicken embryos. In 1938, there was the first intralaboratory infection caused by the newly established pathogen. Rolla Dyer was infected with the Nine Mile agent in Hamilton, Montana. After inoculation of guinea pigs with his blood, febrile illness was reproduced. In the spleen of the test animals were found rickettsiae [174]. The identity of the microorganism isolated from the blood of Dyer and Nine Mile agent was confirmed and cross-immunity between the two isolates was established. Later Dyer [175] himself proved the identity of the American strain Nine Mile (R. diaporica) with Australian R. burnetii. The current name of the agent of Q fever, Coxiella burnetii, was adopted in 1948 at the suggestion of Philip [398] as a credit to Cox and Burnet in the isolation and characterization of new microbial species.

In Europe, Q fever was established for the first time in the Balkans during World War II (1941) as a febrile illness among the German troops, which gained wide popularity under the name "Balkan flu" [98, 120, 162, 258, 259]. Initially, the agent of the disease was isolated from the sputum and blood of human patients after the inoculation of guinea pigs, and without its identification [120, 258]. Later, the identity of the infectious agent of the "Balkan flu" with C. burnetii [143, 144] was established in laboratories in the United States. When studying the sources of infection for humans, Camino Petros [120, 121] found serologically positive for C. burnetii goats in mainland Greece. At the end of the Second World War, there were several outbreaks of Q fever among American and British troops, located in Italy, Corsica, and Greece [145, 186, 440]. These first reports for Q fever cases in Australia, the US, and Europe caused a great interest among scientists from many countries. Literature shows that Q fever has been detected in at least 51 countries [85, 269].

In other countries of the Balkan region, research began to focus on the problem in 1947 when Romania established the disease in people with a probable source of infection, namely sheep [142]. In Turkey, during 1948–1950, isolated Coxiella was established in individuals with a typical pneumonia, and 15.5% of the surveyed 213 sheep were serologically positive for Q fever [389, 390]. In 1949–1951, in Yugoslavia, the first cases of the disease were revealed [356, 368].

In Bulgaria, two clinical cases of Q fever in humans, also confirmed serologically abroad, were discovered in 1949 by Mitov. Pioneer studies in animals were made by Angelov, in 1951 [5]. The team of the same authors conducted serological and virological tests in the period 1951–1957 and found Q fever in goats and cows mainly in Plovdiv County [6–9]. Later, several groups of researchers conducted animal studies from different regions of the country: Ognianov, 1955–1957 [43], in the town of Peshera, found 12% seropositive goats out of 285 tested; and the reports for the findings of Serbezov in the period 1959–1960 were reported in [55, 57]. In the Eastern Rhodopes, part of the settlements in the Balkan Mountains, northeastern Bulgaria with Dobrudzha, Southeast Bulgaria in Strandja, 1036 animals were tested serologically, out of which 243 (23.1%) were positive, with 183 (23.4%) of 692 sheep, 23 (16.3%) of 141 goats, 36 (21%) of 171 cattle; Gelev, 1961 [16], in Northeastern Bulgaria, found seropositive sheep, goats, and cows in 14.5% of 214 surveyed in seven farms; Nikolov, 1961 [40], tested serologically

2356 animals in the region of Sofia, Pernik, Blagoevgrad, Kyustendil, Vidin, Mihailovgrad, Dobrich, Varna, Stara Zagora, and Haskovo districts; 86 of them were positive for coxiellosis with 30 cattle (5.4%), 26 sheep (6.4%) 7 roses (7.3%), 12 horses (2.3%), 8 pigs (2.38%), and 3 dogs (8.1 %); Pandarov, 1963–1968 [47, 48], investigated flocks of 15 sheep farms in Ruse and Haskovo districts and found antibodies against C. burnetii in 591 animals (18.2%). The percentage of positive seroreagents vary considerably, the highest in the Chilnov (Rousse), 83.1%, where during lambing campaign broke outbreak, 74 people were affected, mainly breeders. In the early seventies, studies on the incidence of Q fever among sheep and goats in northeastern Bulgaria were made by Genchev et al. [17, 19].

In the extended period (1977–2012), S. P. Martinov carried out large-scale complex studies on Coxiella burnetii and Q fever in animals throughout the country [586–598]. In this period, a number of epidemic outbreaks of the disease occurred among the population, which will be considered separately.

According to Mitov et al. [36], until 1964, 13 strains of C. burnetii had been isolated in Bulgaria: 7 from the blood of humans, 1 from blood of 1 goat, 1 of milk of a cow, 2 from internal parenchymal organs of hamsters, and 2 of ticks. Data for isolation of C. burnetii of small ruminants and some wild species are exposed in other publications [7, 43–45, 47, 48, 66].

The first reports of the existence of natural foci of Q fever in Bulgaria are found in [17, 41, 42, 57, 66]. The relationship between the C. burnetii infection in animals—mainly sheep and human disease—has been shown in a number of sporadic cases or outbreaks, where the inhalatory route of infection is key [8, 13, 27, 29, 34, 52, 60, 62–64]. Observations in these early Bulgarian publications were confirmed, expanded, and interpreted on advanced level in a number of articles in the 1990s of twentieth century and in the first decade of the twenty-first century [2, 3, 39, 72, 471, 597–600].

2

Etiology and Taxonomy

Coxiella burnetii is an obligate intracellular microorganism with unique genotypic and phenotypic characteristics that make its current taxonomy uncertain and temporary [248, 387]. This contributes to the diversity of qualities and properties of the whole group rickettsial agents united in the order Rickettsiales, family Rickettsiaceae, and branch (tribe) Rickettsiae with three genera: Rickettsia, Rochalimaea, and Coxiella [548]. The criteria for the classification of Rickettsia and other possible related microorganisms belong to the following groups [124].

- *Genotypic*: (a) chain sequence of bases of DNA, (b) ribosomal RNA, (c) migration patterns of protein, and (d) antigenic characteristics.
- *Phenotype*: (a) morphology, (b) structure of the cell wall, (c) staring properties in different techniques including Gram technology, (d) metabolism of L-glutamine, and (e) generating ATP [437].
- *Pathogenetic*: (a) depending on the cell, (c) depending on the vectors, and (c) the relationship of the organism with the cell, (d) type of host cell (lymphoid, myeloid, etc.), (e) cytoplasmatic or nuclear environment, (f) free or inside the vacuole location, (g) preference for replication in vacuoles—phagosomes or phagolysosomas, (h) tolerant or lethal effect, and i) simple division or development cycle.

Difficulties and problems in the classification of rickettsial agents are discussed in detail at Weiss and Moulder [548] and Weisburg et al. [550]. These authors found genotypic relationship but minimal phenotypic proximity between genus Rickettsia and genus Coxiella emphasizing strict characteristic, the unique properties of Coxiella. The disadvantages of this classification are obvious and due to the inclusion of genus Rochalimaea in the branch of the family Rickettsiaceae. Indeed, the latter is genetically similar to Rickettsia in DNA hybridization, but there are fundamental phenotype differences; it is the only member of the order Rickettsiales which can be cultivated in artificial culture media [548, 550].

Much of the above criteria were considered and taken into account when trying to classify rickettsial agents. Some genotypic indicators are used to construct genetic trees—dendrograms [77, 124, 154, 434, 536, 550]. According to Neron [240], similar dendrogram reflects the alleged evolutionary lines of development with the appropriate divergence, but although they are intellectually exciting, they are not always readily obvious.

Currently, based on phylogenetic studies performed mainly by comparative 16S ribosomal RNA sequence analysis, the only genus Coxiella with the only representative C. burnetii is placed in a gamma-Proteobacteria subdivision [490, 549, 550, 558]. The most closely related microorganisms are referred to the genera Legionella, Francisella, and Rickettsiella. Changes in classification affect the other two genera of the previous branch Rickettsiae: Rickettsia is attributed to alpha-1 subgroup of Proteobacteria, and Rochalimaea goes to alpha-2 subgroup of Proteobacteria (within the genus Bartonella, family Bartonellaceae). Phylogenetic tree was constructed using the 16S RNA gene obtained by the method chains "neighborjoining". In this new classification, uncertainties and controversial facts can also be found. Identified as the closest to the Coxiella microorganism, the Legionella is a facultative intracellular parasite with a different growth and development. It has the ability to survive and multiply intracellularly, and the Legionella infection is manifested by a different clinical picture.

Coxiella burnetii is developed and replicated only in the intracellular environment of the host, with the particularity that there is spore-like form of the agent, which is highly resistant to heat, drying, pressure, number of standard antiseptics and the like [557]. This allows the C. burnetii to be kept in the environment for extended periods. McCaul et al. [348] report that spore-like form of C. burnetii can be observed in human tissue in endocarditis. Typical are the small size of the pathogen and the formation of shapes capable of passing through filters. The structure and organization of the Coxiella cells resemble in some respects those of Gram-negative bacteria. Although the membrane of the agent is similar to that of the said bacteria, it is not normally stained by Gram. For staining of light microscopy preparations from clinical or laboratory materials and experimental cultures, the most commonly used methods of Macchiavello [307], Stamp et al. [487], Gimenez [208], and Zdrodovskiy and Golinevich [30] were followed.

The size of the genome of C. burnetii varies within wide limits between different strains, from 1.5 to 2.4 Mb [556]. In the American Nine Mile strain, it is 2.1 Mb. Chen et al. [130] indicated that the site of replication of DNA is into the chromosome of C. burnetii. This statement, however, at this stage is rather hypothetical [504]. According to Willems et al. [556], the failure of the attempts to detect the site of replication of the genome of C. burnetii by standard methods seems to be due to the likelihood that C. burnetii to have linear and not circular chromosomes. The latter in turn would determine the absence of a conventional two-way replication [504]. Contributing to the issue of creating a genetic map, respectively comparing the localization of genes and genetic organization of the various isolates of C. burnetii was made by Willems et al. [555, 556]. The authors constructed a physical map of the Nine Mile strain in the phase I, which can serve as a basis for the elaboration of the above-mentioned genetic map. Electrophoretically (pulsed-field gel electrophoresis—PFGE), after restriction of the total DNA with *NotI*, 25 fragments of DNA were separated. Attempts to clone the ends of the 240– and 7.3 kb fragment of C. burnetii (Nine Mile I phase) in a compatible vector, after removal of the said fragments from the gel of PFGE and enzymatic digestion, were unsuccessful. These experiments support the assumption of linear chromosomes [555, 556]. Heinzen et al. [234] also used PFGE to differentiate C. burnetii strains. Several authors turned their attention to cloning and expression of chromosomal genes of C. burnetii in E. coli. Heinzen and Mallavia [233] reported the gene gltA (citrate synthase gene), Vodkin and Williams [539] for two genes (heat shock protein gene) with names corresponding 14 kDa and htpB-62 kDa, and Hendrix et al. [238] 27 kDa surface antigen gene. The chronology of these studies continues with Hoover and Williams [247], which characterize gene pyrB (aspartate carbamoyl transferase gene); Heinzen et al. [235], superoxide dismutase gene; Mo and Mallavia [359], sensor-like protein gene; Zuber et al. [582, 583], gene dnaJ (heat shock protein gene) and gene mucZ (capsule induction protein gene); and Willems et al. [555], gene algC (phosphomannomutase gene).

Interesting is the question of genomic variations between different strains of C. burnetii [239, 511, 555, 582, 583]. These are less pronounced when using the method of DNA–DNA hybridization [537]. The application of the restriction analysis, RFLP (restriction fragment length polymorphism analysis), on the DNA of 38 isolates led to the formation of 6 genomic groups, from I to VI [239]. Subsequently, using PFGE and *NotI*, the study of 80 strains of C. burnetii, originating from different geographical regions of the world, identified 16 new genomic groups [511, 555]. In the dendrogram constructed by an UPGMA method, including strains of animal (sheep, goats) and human (acute and chronic Q fever) origin, strong genetic heterogeneity (7–58%) was established between different groups [511, 555]. Zhang et al. [580] differentiate C. burnetii by a sequence analysis of the gene (com 1) encoding 27-kDa outer membrane protein. They found that strains derived from ticks, cattle, and people with acute Q fever at the molecular level are different from the strains isolated from persons with chronic Q fever.

Another feature of the agent of Q fever is the presence of plasmids. To a part of them they are found to be associated with resistance to antibiotics and pathogenicity, and other functions are unclear yet. In the studies of strains isolated from patients with acute and chronic forms, and including the plasmid composition profile of the DNA fragments in the restriction endonuclease mapping, lipopolysaccharide profile, etc., differences were established [209, 255, 343, 435]. First, Samuel et al. [454] described the plasmid QpH1 (36 kb, 1–3 copies per cell) isolated from C. burnetii in phase I. Parallels between the plasmid and the type of disease caused by this pathogen [455] were defined. Thiele et al. [512] determined the complete nucleotide sequence of QpH1. This plasmid was found in genomic groups I, II, and III of C. burnetii. Another plasmid, QpRS (39-kb), was established in genomic Group IV [312]. QpRS was initially characterized in strain C. burnetii, isolated from aborted fetuses of a goat and later in human isolates from cases of endocarditis. Mallavia [312] reported the isolation of plasmid QpRS (42-kb) from wild rodents, which is located in Coxiella isolates of genomic-VI group. In another case of endocarditis in humans was detected plasmid QpDV, 33-kb [534]. In some isolates of C. burnetii (endocarditis), relating to the genomic V group, plasmids are missing, but were detected chromosomal DNA sequences exhibiting homology with the plasmid QpRS [460, 555].

3

Morphological Features, Development Cycle, and Interaction with Cells

Initial knowledge of the morphology of C. burnetii is based on observations by an ordinary light microscope. Also, Burnet and Freeman [117] and Cox [149] indicated the presence of a polymorphism, stating that the most typical are small coccus-like forms with an approximate size of 500 nm (length) and 250 nm (diameter). Rod-like and spherical forms of different sizes, filamentary formations, and chains were also observed. Common findings are aggregations of polymorphic Coxiella bodies in the form of nests into the cytoplasm of the infected cells. Extracellularly located cells of C. burnetii were observed in [30]. The microscopic finding is nuanced in color terms, depending on the method used for the staining preparations. For example, in the application of different methods in the microscopic preparations from cotyledons or placental tissue of aborted animals, a large number of thin, pink-, red-, ruby red-, or violet-colored bacillus-like or pleomorphic cells on a blue or green background are observed (×500). Microscopy is often accompanied by difficulties and experience required due to the small size of the agent—300–1500 nm in length and about 250 nm in width [409]. The light microscopy methods for the morphological evidence of C. burnetii continue to be part of the diagnostic approaches in a number of laboratories [294, 343, 409]. They do not have the capacity, however, for the study of the fine morphology and mode of development of the causative agent.

With the advent of electron microscopy methods (EM), a new light on the morphology and morphogenesis of C. burnetii was shed [1, 24, 50, 76, 152, 211, 281, 358, 377, 436, 543, 552]. Cited earlier studies substantiate the first knowledge of the ultra structure of the agent and its development cycle. Most authors take the view that C. burnetii reproduces in a similar mechanism like bacteria and undergoes a binary division. In contrast to this view, Cordova et al. [281] described the development cycle comprising eclipse, the occurrence of the fine granules; a series of larger and large intermediate-forms without an outer membrane (matrix) and the presence of a compact electron dense entities and small granular particles; formation of pleomorphic cells; and lack of double division.

Other studies based on EM and biochemical techniques indicate that C. burnetii has a complex intracellular development cycle and interaction with the host.

Although C. burnetii proliferate *in vitro* and *in vivo* in different cells, previous studies have shown that primary infection affects two types— monocytes and macrophages [113, 216, 286, 322, 326]. According to Mege et al. [354], the entry of C. burnetii in these cells involves specific receptors: LRI (leucocyte response integrin) and IAP (integrin-associated protein) to the agent in first phase and CR3 for the second phase. After entering the cell, the agent is located in phagosomes, which recently have been merged with lysosomes. New formations are designated as phagolysosomas. The last progressive merge in a large vacuole with unique features is reported in [218]. Heinzen et al. [236] performed biochemical studies and concluded the lysosomal origin of vacuoles containing C. burnetii, based on the discovery of lysosomal glycoproteins and markers (proton-ATPase, cathepsin D, and acid phosphatase) in them. Essential condition for the reproduction of the pathogen in vacuoles is the presence of an acidic environment—pH 4.7–5.2 [70, 340, 417]. Metabolic processes in C. burnetii, including the synthesis of nucleic acids and amino acids, are dependent on the acidity [131, 237, 584]. Maurin et al. [340] indicated that phagolysosomas of cells infected with Coxiella maintain an acidic pH during the persistent infection. Hackstadt and Williams [219] reported a pH-dependence of glutamate transport system of C. burnetii.

In pleomorphic population of C. burnetii in phagolysosomas of infected yolk sacs of chicken embryos, McCaul and Williams [345] distinguish two main categories cells: large—"large cell variant" (LCV) and small—"small cell variant" (SCV). Both morphological forms differ in their ultrastructure, peptidoglycan content, and resistance to osmotic pressure [73, 344, 345]. Great cells (LCV) reach 2 μm length. They exhibit a high degree of polymorphism, but rounded cells coated with a two-layer cell wall with outer and inner (cytoplasmic) membrane separated by a periplasmic space are dominating. The internal content consists of granular or granular–fibrillar plasma and scattered nucleoid segments. LSV contains less peptidoglycan in the cell wall. Metabolically, the more active intracellularly located are the LSV. They have the ability to produce spore-like forms, characterized by greater resistance [346, 347]. Initially, the spore-like form has endogenous localization and after further development becomes metabolically inactive category SCV.

The latter leaves the host cell after lysis of the cell wall. This already extracellular form of C. burnetii is resistant to drying, chemical effects, disinfection in normal concentrations, high or low values of hydrogen-ion concentration, and ultraviolet radiation [85, 466].

The small cells (SCVs) are rod-shaped, with extremely thick cell walls and electron-dense nucleoids. Their size is approximately 450 nm in length and 200 nm diameter. Their cell wall is double layered with periplasmic space filled with dense matter. It is believed that SCVs are metabolically inactive. They are resistant to osmotic pressure and constitute the extracellular form of C. burnetii. Small cells are attached to the membrane of phagocytic cells, enter the phagolysosomas by fusing, and after acid activation of the SCVs metabolism may lead to conversion into LCVs [346, 347]. For both morphological forms, LCVs and SCVs, a binary division is observed.

The factors of organic, chemical, or physical nature, which can cause similar sporulation process in C. burnetii, are not known yet at the time of writing of this book. The duration of the replication of C. burnetii is significant—about 20 h [343]. According to Winstead, [560] this period is twice small ($\sim$10 h). Some authors associate the ability of C. burnetii to induce persistent, chronic, and latent infections with the slow pace of intracellular multiplication of the agent [87, 88, 126, 494, 510]. Seshadri et al. [472] reported deciphering the complete genome sequence of C. burnetii—Nine Mile strain in the phase I (RSA493), 1,995,275-bp genome. The authors found that the pathogen is highly adapted to existence in eukaryotic phagolysosomas. Genomic analysis uncovered many genes with a potential role in adhesion, invasion, intracellular movement, and modulation of host cell. Although the lifestyle and parasitism of C. burnetii resembles the one identified in Rickettsia and Chlamydia, the genomic architecture of C. burnetii differs considerably in terms of mobile elements, the degree of genomic reduction, the metabolic properties, and the transport capabilities [472].

4

Ultrastructure of Coxiella burnetii and Electron Microscopic Diagnosis of Q Fever

Chapter 3 presented some general morphological and morphogenetic details of C. burnetii as a prerequisite for the clarification of questions about the way of development and interaction of the etiological agent with the cells.

In this section, the emphasis is on specific morphological features of the agent in clinical, pathological, and experimental materials of different origin, mainly in view of the morphological diagnosis, as reflected in our considerable personal experience (Popov and Martinov [601, 602], Martinov and Popov [586, 590, 603], where along with data, also an idea on the reproduction of the pathogen and pathogenesis of the disease was given.

4.1 Light Microscopic Morphology of C. burnetii

The microscopic demonstration of C. burnetii required impression preparations and smears from different clinical, pathological, and experimental materials: placentas, cotyledons, uterus, spleen, livers, lungs, and other parenchymal organs, fibrinous deposits, peritoneal exudates, yolk sacs of chicken embryos, etc. The preparations were stained according to the methods of Stamp, Macchiavello, Zdrodovskiy, or Golinevich.

Infected placentas and cotyledons in abortions, stillbirths, and similar conditions are very suitable for this type of study. We observed C. burnetii in the form of pleomorphic cells—spherical, oval, rod-like, or intermediate with ruby red color (Figure 4.1). Coxiellas are located intracellularly in the cytoplasm, singly or in small, medium, and large groups. There are also extracellularly located Coxiella cells, mostly single or small groups. The amounts of the agent are not uniform, from 250 to 450 nm and from 500 to 800 nm and larger.

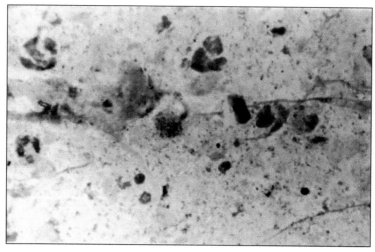

Figure 4.1   Light microscopy. Coxiella burnetii in smear from cotyledons of aborted sheep with spontaneous Q fever. Staining by Macchiavello. Magnification 1000X.

Less suitable for light microscopic examination are the fetal parenchymal organs where apparently due to weaker accumulation of the agent, typically single coxiellas are observed and rarely conglomerates of these.

In the isolation of C. burnetii in guinea pigs and white mice after intraperitoneal infection, the agent settled in large amounts in the preparations from the spleen (see Figure 4.2). After staining using the methods of Zdrodovskiy and Golinevich, we observed polymorphic bright red cells on a blue background with a light fawn hue in the cytoplasm of splenic cells.

With the efficient infection and its development in microscopic preparations of the outer and the cut surface of the liver, we could detect coxiellas with varying degrees of accumulation and a similar location within and outside the cells.

Located on the surfaces of the spleen, liver, and other parenchymal organs of mice and guinea pigs, off-white fibrinous taxed surfaces contain significant or copious amounts of the agent as a predominant rod forms. In preparations of the peritoneal fluid of experimental animals, coxiellas in different concentrations could be found.

Infection with the agent of Q fever led to very typical pathological changes in chick embryos and especially in their yolk sacs, which is a useful indicator of the infection. In the corresponding stained preparations or impression smears were observed punctate, oval, and coccus-like coxiellas colored with different

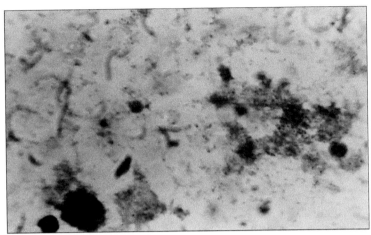

Figure 4.2 Light microscopy. Coxiella burnetii in smear spleen of guinea pigs infected intraperitoneally with a suspension of parenchymal organs of stillborn calf. Staining by Zdrodovskiy. Magnification 1000X.

shades of red or purple and located in the cytoplasm of endodermal cells in the form of inclusions or diffusely available. Figure 4.3 shows a light microscopy preparation (Stamp) from yolk sac of a chicken embryo infected with milky exudation of sheep affected by Coxiella-induced mastitis.

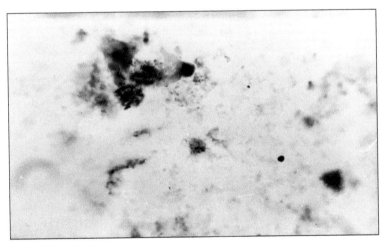

Figure 4.3 Light microscopy. Coxiella burnetii mastitis in sheep. Coxiellas in impression preparation from yolk sac of chicken embryo infected with milky exudation. Staining by Stamp. Magnification 1000X.

4.2 Ultrastructure of the Agent and Electron Microscopic Diagnosis of the Disease

4.2.1 Placentas in Abortions and Births of Dead and Unviable Offspring

In negatively stained preparations from placentas and cotyledons, C. burnetii was found in high concentrations, up to 10^9/ml suspension. The bodies were rod-shaped or oval and have a size of 200–1500 nm.

In ultrathin sections of placental tissue and cotyledons with macroscopically visible changes, the pathogen is located intracellularly and extracellularly. The intracellular location of C. burnetii is characterized by the presence of the agent into the cytoplasm of the infected epithelial cells in the form of solid inclusions consisting of multiple Coxiella cells. There are also coxiellas as smaller groups or individually arranged. In some cases, the agent is situated in electron transparent vacuoles or freely next to the cell wall of the infected cells (Figures 4.4 and 4.5).

The shape of the C. burnetii is rod-like with an average size of 200–900 nm, and oval or spherical in two types of cells: small with diameter 220–280 nm, and large with diameter 700 nm (Figures 4.6–4.8). Rod and spherical forms occur equally often.

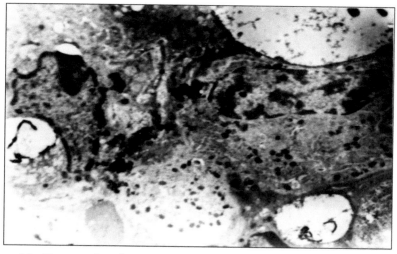

Figure 4.4 Electron micro photograph. Spontaneous abortion in cattle affected by Q fever. Ultrathin section of placenta. Multiple coxiellas located in groups or individually in the cytoplasm of epithelial cells. Magnification 10,000X.

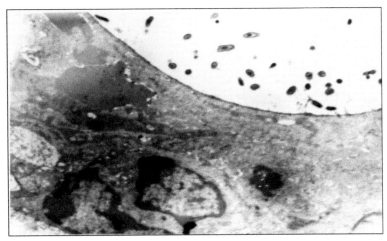

Figure 4.5 Electron micro photograph. Q fever in cattle. Ultrathin sections from aborted placenta. A large vacuole in the cytoplasm of epithelial cells, comprising oval, spherical, and rod-like cells of C. burnetii. Magnification 10,000X.

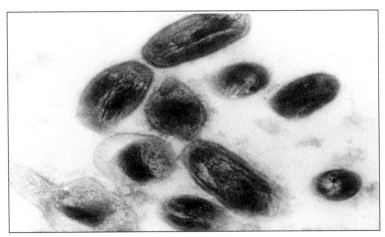

Figure 4.6 Electron micro photograph. Experimental Q fever in sheep. Ultrathin section of placenta after a stillbirth. Multiple coxiellas in the cytoplasm of epithelial cells. Magnification 80,000X.

Surveys show that only about half of the agent's cells in the Coxiella burnetii population have normal cell structure. In morphologically intact cells of the pathogen, a three-layer membrane and inner contents can be

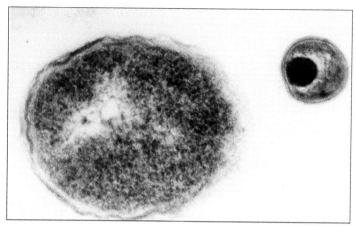

Figure 4.7   Electron micro photograph. Experimental Q fever in sheep. Ultrathin section of placenta after the birth of a non-viable lamb. Large and small spherical Coxiella cells. Magnification 150,000X.

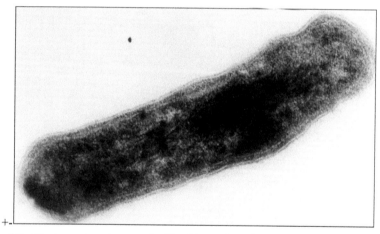

Figure 4.8   Electron microphotograph. Ultrathin sections from the cotyledon after the still-birth of ewe-lamb experimentally infected with C. burnetii. Rod-like cell of the pathogen. Magnification 100,000X.

seen, consisting of a granular cytoplasm type and the availability of ribosomes and osmiophilic filiform-like nucleoid, most often centrally located. In other Coxiella, the cell cytoplasm is fibrillar, and the nucleoid is not compact, but diffusely scattered. Approximately half of the cells, both rod-like and spherical, are in a state of degeneration. Cell membranes, cytoplasm, and nucleoid are equally affected by degenerative changes. For example, lysis and homogenization of the cell membrane, homogenization, and vacuolization of the cytoplasm, and homogenization and congestion of the nucleoid are observed. In the Coxiella population, along with destroyed cells and cells with normal morphology, dividing cells are seen. The rod-like cells divide into two parts by constriction.

The placenta and cotyledons cells themselves, infected with C. burnetii, display pathological changes. In the cytoplasm, vacuolization was observed, as well as the occurrence of primary and secondary lysosomes and large lipid droplets. In the central regions of most of the mitochondria, vacuolization is present, which gradually would cover the entire structure and would lead to its total destruction. The ducts of the endoplasmic reticulum are widened and some of them have single C. burnetii cells (Figures. 4.4 and 4.9). The nuclei of the infected epithelial cells are pyknotic. The amount of heterochromatin is reduced, and the nuclear pores are expanded. In a more highly injured cells, disintegrating swelling of nuclei is found, wherein the combined nuclear edema with grouped chromatin at several sites in the nuclear membrane (see Figures 4.4 and 4.9).

There are also completely destroyed epithelial cells, of which there are only masses of detritus, and many coxiellas. There are also epithelial cells turned into giant vacuoles filled with freely arranged Coxiella cells (Figures 4.4 and 4.5).

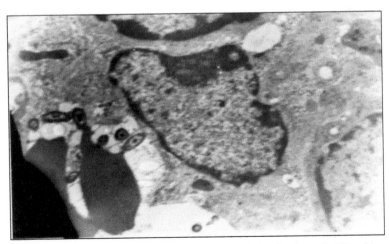

Figure 4.9 Electron micro photograph. Coxiella burnetii abortion in cattle. Ultrathin section of placenta. Abnormal changes in the cytoplasm and nucleus. Coxiellas clinging to the erythrocyte. Magnification 20,000X.

We will note that C. burnetii was also established in a direct electron microscopy of pathological material from animals with abortion and related conditions having mixed infections—Coxiella and Chlamydia, Coxiella and Salmonella, or other bacterial agents.

4.2.2 Blood

In the blood vessels of the placenta, abundant coxiellas were observed, most of which are positioned freely in the plasma of the blood flow among the formed blood elements (Figures 4.10–4.13). The displayed electron microphotographs demonstrate spherical and oval coxiella cells with clearly distinguishable membranes, diffuse nucleoids, and ribosomes. In single coxiellas, the nucleoid has a greater degree of compactness and a centrally or slightly off-center location.

The causative agent in the form of inclusions and single cells was detected in the cytoplasm of the endothelial cells of the blood capillaries wherein abnormalities were observed—analogous to those described above for the epithelial cells of the placenta (Figure 4.15). The probable sequence of penetration of the pathogen into the bloodstream includes relevant epithelial

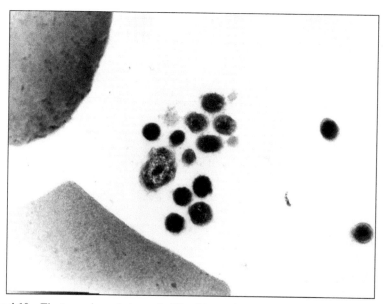

Figure 4.10   Electron micro photograph. Ultrathin sections from the blood of aborted sheep. Coxiella burnetii in the plasma of the blood flow. Magnification 20,000X.

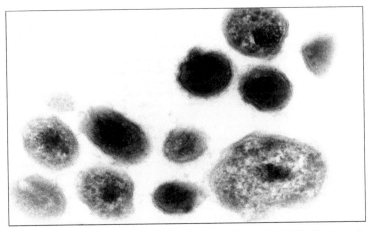

Figure 4.11 Electron micro photograph. A fragment of Figure 4.11. Morphology of spherical and oval coxiellas in the bloodstream. Magnification 50,000X.

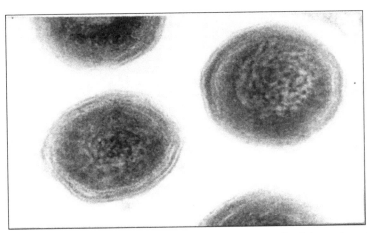

Figure 4.12 Electron micro photograph. Ultrathin section of blood from aborted sheep. Spherical cells of C. burnetii with membrane, ribosomes, and diffuse nucleoid. Magnification 220,000X.

cells depending on the front door of infection (respiratory or gastrointestinal tract), destruction of epithelial cells, passing in endothelial cells and their infection, and penetration from endothelial cells in the plasma of the blood flow. The presence of coxiellas that adhere to the cell membrane of endothelial cells gives rise to the assumption that C. burnetii pass into the bloodstream through the endothelium.

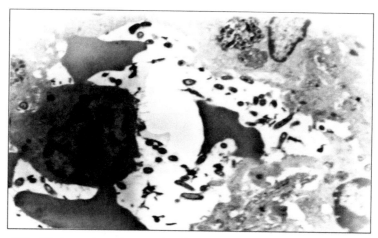

Figure 4.13   Electron micro photograph. Ultrathin sections from placenta of aborted cow with Q fever. There are Coxiella cells in the bloodstream and coxiellas connected with bridges to erythrocytes. Magnification 10,000X.

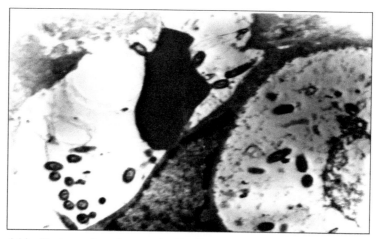

Figure 4.14   Electron micro photograph. Abortion at Q fever in cows. Ultrathin section of placenta. Location of C. burnetii is as follows: (a) freely located in the plasma of blood flow, (b) closely adjacent or connected with bridges to the erythrocyte, and (c) in the large cytoplasmic vacuole of an epithelial cell. Magnification 14,000X.

Another part of the coxiellas is clinging tightly or connected with a thin bridge to erythrocytes (see Figures 4.9, 4.13 and 4.14).

A particularly interesting finding was the presence of C. burnetii within erythrocytes, leading to their damage and destruction. Some of the erythrocytes are almost entirely vacuolated, and in others can be seen in several smaller vacuoles (Figure 4.15).

Leukocytes infected with the pathogen are also observed. Figure 4.16 has demonstrated morphological findings in an ultrathin section of sheep placenta,

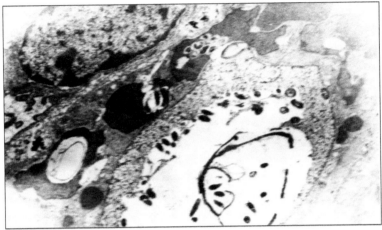

Figure 4.15 Electron micro photograph. Q fever in cattle. Ultrathin sections from aborted placenta. Presence of C. burnetii within red blood cells and in the cytoplasm of endothelial cells. Magnification 12,000X.

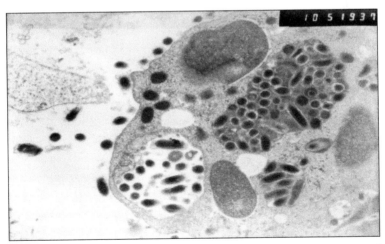

Figure 4.16 Direct electron microscopy. Inclusions of Coxiella burnetii in the cytoplasm of a leucocyte from ultrathin sections prepared of cotyledon from miscarriage sheep. The agent is also seen in the space outside the cell. Magnification 25,000X.

aborted at the fifth month. In the infected polymorphonuclear leukocyte, visible large inclusions of C. burnetii, smaller groups or single cells of the agent, were found in the cytoplasm. Some of these are located at the periphery of the cell wall, from which leave the cell by two possible mechanisms: after lysis of the section in which they are located, by exocytosis. In the cytoplasm of infected leukocyte, vacuoles, some of which are filled with many coxiellas, were detected. Outside the cell, the agent is mainly localized in close proximity with the cell wall, forming a "festoon" of C. burnetii bodies (Figure 4.16). Fragments of those are shown in Figure 4.16 and inclusions are presented in Figures 4.17 and 4.18, which show the typical ultrastructure of C. burnetii.

4.2.3 Chicken Embryos

Electron microscope indication of the strains C. burnetii, isolated in yolk sacs of chicken embryos from placentas, fetuses, milk secretions, and other materials, is possible even at the primary infection. The pathogen in negatively stained suspensions is rod-shaped or oval and has a size of 200–1500 nm. In ultrathin sections, the coxiellas have a three-layer membrane and grain-like and fibrous inner content with diffuse nucleus. Their shape is oval or rod-like. The size varies from 200 to 800–1500 nm (Figures 4.19–4.21).

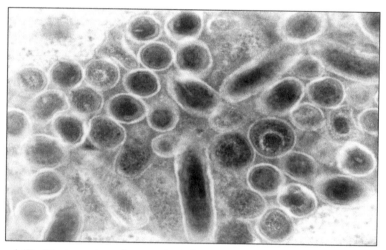

Figure 4.17   Electron micro photograph. A fragment of a C. burnetii inclusion, shown in Figure 4.16. Morphology of the rod-shaped and spherical forms of the pathogen. Magnification 50,000X.

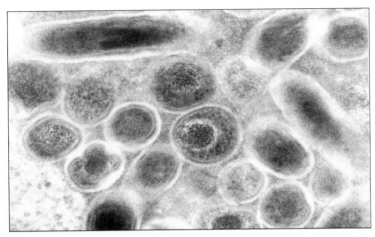

Figure 4.18 Electron micro photograph. Fragment of the inclusion shown in Figure 4.17. Typical ultrastructure of C. burnetii: a three-layer membrane, ribosomes, and diffuse nucleoid. Magnification 70,000X.

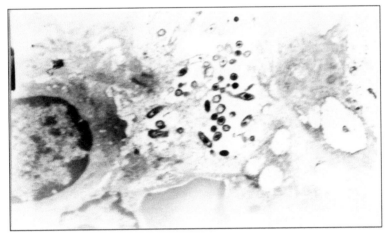

Figure 4.19 Electron micro photograph. Ultrathin section of yolk sac of chicken embryo infected with C. burnetii—Tutrakan-1 strain isolated from milk of a sheep with mastitis. Multiple Coxiella cells. Magnification 12,000X.

4.2.4 White Mice

Figure 4.22 shows the EM indication of C. burnetii in ultrathin sections from the brains of newborn white mice, after the intracerebral inoculation with the blood of sheep affected by pneumonia. There is massive compactin clusion of

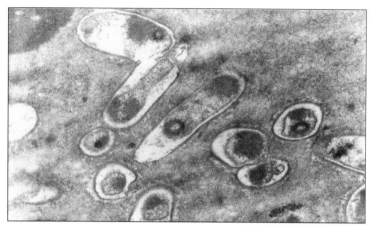

Figure 4.20   Electron micro photograph. Ultrathin section of yolk sac of chicken embryo infected with C. burnetii—Sapareva banya strain isolated from internal parenchymal organs of Faborted bovine FETUS. Polymorphic Coxiella bodies. Magnification 40,000X.

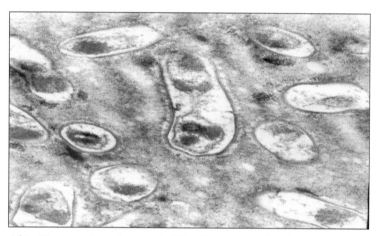

Figure 4.21   Electron micro photograph. Ultrathin sections of yolk sac of chicken embryo infected with C. burnetii—Bistrica strain isolated from cotyledons of sheep after the birth of a non-viable lamb. Oval and rod-like forms of C. burnetii. Magnification 100,000X.

the agent located into the cytoplasm and consisting of multiple polymorphic Coxiella cells, tightly spaced together. Dominating are the oval and rod-like forms. Part of the bodies are arranged in small groups in the cellcytoplasm. A similar morphological picture was observed in strain C. burnetii, isolated in the brain of a newborn white mice after inoculation of blood from hare.

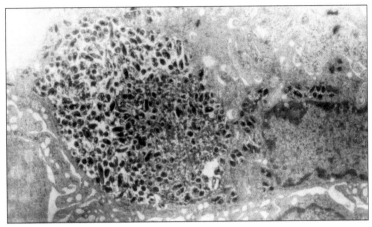

Figure 4.22 Electron micro photograph. Ultrathin sections from the brain of newborn white mice infected intracerebrally with blood of sheep with pneumonia. C. burnetii in the form of massive inclusion in the cytoplasm. Magnification 12,000X.

In EM of parenchymal organs from young white mice infected intraperitoneally with a suspension of fetal organs and placentas from sheep and goats with Q fever, the richest finds of the agent were found in the spleen. Coxiella burnetii is detected in the form of inclusions of multiple coxiellas located in the cytoplasm of cells in the spleen (Figure 4.23).

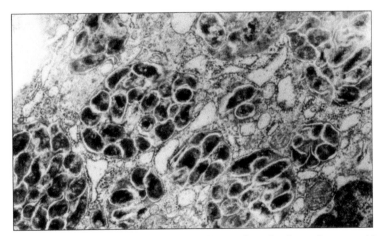

Figure 4.23 Electron micro photograph. Ultrathin section of spleen of young white mice infected intraperitoneally with fetal and placental material from sheep sick of Q fever. Multiple inclusions of pleomorphic coxiell as in the cytoplasm of the spleen cells. Magnification 40,000X.

4.2.5 Guinea Pigs

We investigated parenchymal organs from guinea pigs that had been infected with intraperitoneal inoculations of suspensions from placentas, organs of fetuses, modified milk, and other secretions. The infectious agent was observed in the cytoplasm of the lymphoid cells of the spleen in the form of large conglomerates /inclusions/ of a plurality of C. burnetii bodies with rod-like, oval, spherical, or irregular shape at various stages of the reproductive cycle (Figures 4.24 and 4.25). Part of the C. burnetii bodies showed degenerative changes.

From the presented results, it can be seen that the ultrastructural features and intracellular multiplication of C. burnetii are typical for the presented species. They allow precise morphological indicators and identification of C. burnetii and the exact differentiation of bacteria and chlamydia. The electron diagnosis of Q fever is fast and highly efficient [601, 602]. It gives ample opportunities to detect the pathogen in clinical, pathological, and experimental materials and to identify pathological changes in infected these cells and tissues.

We conducted studies on the morphology of C. burnetii, which were based on two main approaches—the use of conventional light microscopy (LM) and the methods of modern transmissible electron microscopy.

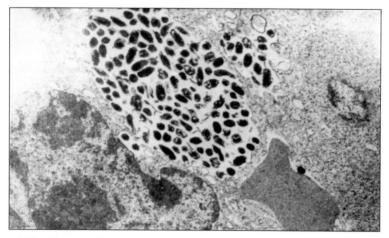

Figure 4.24   Electron micro photograph. Ultrathin section of guinea pig spleen infected intraperitoneally with a suspension of placenta and fetal parenchymal organs in C. burnetii abortion in sheep. Massive conglomerate of polymorphous cells of C. burnetii. Magnification 25,000X.

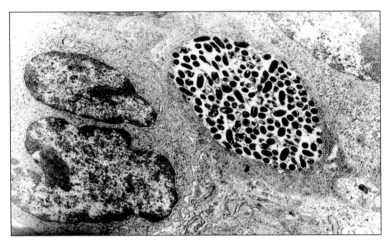

Figure 4.25 Electron micro photograph. Ultrathin section of spleen of guinea pig. Intraperitoneal inoculation with a suspension of placenta of aborted sheep. Inclusion consisting of a plurality of coxiellas in the cytoplasm of spleen cells. Magnification 20,000X.

Opportunities for the light microscopic evidence of coxiellas were tested on numerous and diverse clinical, pathological and experimental materials. Years of experience allowed us to use most of the opportunities of the method and to identify the most suitable materials for this kind of research. Such are undoubtedly the cotyledons and placentas at abortion, stillbirths, and other states, in which as a rule, rich deposits of C. burnetii cells were discovered. Structures of the spleen of guinea pigs and white mice infected intraperitoneally with properly selected infectious starting materials also contained large amounts of the agent and could serve as an indicator of its morphology and identification. The findings in the liver, although not so regularly and expressive as in the spleen, served as a source of information about the presence of the pathogen. This also applies to fibrinous taxed on the surface of the spleen, liver, and other parenchymal organ of the said animals. This is not the issue of light microscopic indication of the agent in parenchymal organs of fetuses and adult animals, where the findings are scarce and their LM indication is much more difficult.

In LM preparations with sufficient accumulation of the agent, we observed pleomorphic cells with extracellular and intracellular location manifested by shades of color terms depending on the method of staining. These observations are consistent with the information on the morphology of C. burnetii based on a research with an ordinary light microscope, presented by the pioneers of rickettsiology [2, 4] and confirmed in a number of later studies [30, 44, 47, 57, 294, 343, 409].

When conducting LM studies of preparations of field materials, the acute issue arises for the morphological differential diagnosis with respect to Chlamydia, Brucella, and other bacterial agents [409]. In a significant number of cases, the diagnosis of Q fever in animals based on LM needs to be validated by received positive serological results [409]. In LM observations of preparations from yolk sacs of chicken embryos or cell monolayer, in attempts to isolate the pathogen, the same conditions basically apply: to eliminate the possible presence of other microorganisms, especially bacterial contamination, the accurate differentiation of chlamydia, the sufficient concentration of the agent, often achieved with multiple consecutive passages [409, 343]. In some cases, the preparation of LM-preparations was replaced with a more appropriate immunofluorescence staining.

Generally, the light microscopy methods, regardless of the listed conditions for their adequate use, are useful and can be applied. However, their possibilities for studies on fine morphology and morphogenesis of C. burnetii are extremely limited.

Knowledge of the ultrastructure of coxiellas and the manner of their development is based on the use of electron microscopy [1, 24, 50, 76, 87, 211, 281, 371, 543, 552]. Early studies postulated the perception that C. burnetii in its individual development in the cytoplasm of the host cell, passes through a series of successive morphological changes similar to those seen in bacteria. Other authors [281] described the opposite scheme, which is a kind of virus-like morphogenesis without binary division.

Modern fundamental research based on EM and biochemical techniques contribute to a more complete characterization of the ultrastructural features of C. burnetii and complex intracellular development cycle and interaction with the host [286, 322, 326, 345, 346, 354].

Our electron microscopic observations are the result of research on a large number of clinical, pathological, and experimental materials. Our received data indicate that the populations of coxiellas are complex and polymorphic, and their ultrastructural features allow the exact differentiation from other organisms. Intracellular reproduction of C. burnetii is typical for the species. It is different from the multiplication of bacteria and intracellular morphogenesis of Chlamydia [404].

Electron microscopy has allowed us to make detailed observations of the polymorphism of the population of C. burnetii. The results demonstrate that mature populations of the agent are characterized by a wide variety of cytoplasmic vacuoles, spherical, oval, rod, and intermediate forms, single coxiellas located intracellularly or extracellularly, groups of coxiellas of different sizes in the cytoplasm of infected cells, including in electron transparent vacuoles. The morphological diversity of the C. burnetii population, both in terms of the form of agent and its dimensions, clearly outlines two types of cells—small and large meeting of SCV (small cell variant) and LCV (large cell variant), markings offered by Mc Caul and Williams [345]. The presence of multiple spherical coxiellas with a diameter of 200–300 nm testified that a large part of the population consists of filterable forms. We find it essential that approximately half of the population of C. burnetii hasa normal cell structure, and the rest is affected by degenerative changes. The presence of a degenerative lesions in about 50% of the Coxiella cells is most likely a consequence of the strong impact, respectively lysing effect of the antibodies on the agent as these antibodies are second phase, i.e., antibodies toward the inner antigens found after cell membrane lysis [590].

Noteworthy are the data on the discovery of coxiellas located freely in the plasma of blood flow among the formed blood elements, as well as in infected leukocytes, erythrocytes, and endothelial cells of blood capillaries. The reproduction of C. burnetii in the cytoplasm of polymorphonuclear leukocytes is intensive and is manifested by the presence of numerous cells of the agent arranged singly and in particular in the form of cytoplasmic inclusion, part of which fills the observed vacuoles. Of special interest are the erythrocytes to adhere to their outer surfaces coxiellas, which is the basis for the assumption that such erythrocytes have a transport function of the pathogen in the bloodstream [597]. We will emphasize that our findings for the presence of C. burnetii in erythrocytes, leading to their destruction vacuolization, is the first ever description of this kind in the literature. We observed inclusions of C. burnetii and single cells of the agent in the cytoplasm of endothelial cells in the blood capillaries of the placentas and cytopathic changes in the endothelium, similar to those described in the epithelial cells. Established findings for coxiellas, tightly abutting the inner layer of the cell membrane of endothelial cells, possibly is related to the mechanism of the penetration of the agent from the endothelium (exocytosis or lysis) in the blood stream [597].

Our EM research allowed to obtain new information on the reproduction of Coxiella burnetii, the morphology of the population, and the pathogenesis of Q fever.

The ultrastructural features of C. burnetii and the comparisons between the possibilities of electronic and light microscopy necessitated the need to improve the morphological indication of the pathogen and its identification and development of electronic microscopic diagnosis of Q fever.

- The imperfections and shortcomings of light microscopy are the following: The resolution of light microscope is 200 nm. This means that by theoretically using LM, microorganisms with a size of above 200 nm would be observed. In practice, however, optical microscopes allow for credible monitoring only within the 1000 nm. Therefore, coxiellas with smaller sizes can be found only in the most sophisticated light microscopes and often with considerable difficulties;
- The methods for staining C. burnetii in LM-preparations do not give the same results. There is no perfect method for the LM-staining agent;
- The tested organs and tissues are abundant in particles that have a size of coxiellas and they are a source of artifacts in the LM preparations.

These shortcomings and difficulties show an obvious anachronism to diagnose C. burnetii by light microscopy. In the recent decades, electron microscopy made revolutionary changes in the observations of the micro world. For example, in virology, the electron microscope is the only method for the observation of virus particles, including viruses with larger dimensions, closer to those of the coxiellas. In chlamydiology, the EM-diagnostics developed by us [35, 333, 334, 404] is routinely used. There is clearly a lack of modern EM technology for a more accurate and demonstrative morphological diagnosis in rickettsiology. Therefore, the special literature lacks such reports for the diagnosis of Q fever.

Judging from this situation, we studied the capabilities of the EM methods for the direct detection of C. burnetii in a clinical, pathological, and experimental materials from animals. The results of our studies on ultrathin sections and negatively stained preparations of such materials exhibit a high diagnostic value of the direct EM. A number of positive test results in placentas and aborted fetuses of ruminants were confirmed by the isolation of the pathogen and through serological testing. In the direct EM diagnosis of Q fever abortions, stillbirths, and the delivery of unviable offspring, which often have a mass character and important epidemiological significance, we reached the following conclusions:

- It is necessary to supply primarily placentas and cotyledons, which allows for an easy detection of significant accumulations of C. burnetii as in negatively stained suspensions, and in ultrathin sections for transmission electron microscopy.

- These methods are useful for the study of parenchymal organs (mainly spleen, liver, and lungs) of aborted fetuses and stillborn animals, but as a rule, the pathogen is in lower concentration than in the cotyledons and placental tissue. This may require pre-concentration of the agent by differential centrifugation and super centrifugation.
- A mandatory condition is that the material is fresh. This is particularly important for a parenchymal organ, unlike the placental cotyledons wherein it is possible to detect coxiellas in older materials. The latter explains the greater strength and stability of the tissues against decay processes.
- When using the process of negatively staining, suspensions prepared from 2 g of tissue are sufficient.
- Discovery of C. burnetii is possible in materials from animals with mixed infections—coxiellas and chlamydia, salmonella, listeria, or other bacterial agents. These data are very important, because when there is a mixed infection with bacteria, a normal insulation of the koksieli is impossible due to the bacterial contamination of CE or CC. In a mixed infection with chlamydia, the possible simultaneous isolation of the pathogen requires an EM-morphological differentiation on the basis of typical ultrastructural signs species.
- The detection of C. burnetii is possible in materials from animals with mixed infections—Coxiella and Chlamydia, Salmonella, Listeria, or other bacterial agents. Therefore, in a mixed infection with bacteria, the normal isolation of C. burnetii is impossible due to the bacterial contamination of the chicken embryos and cell cultures. In a mixed infection with chlamydia, the simultaneous isolation of both pathogensis possible, which requires the additional EM morphological differentiation on the basis of their ultrastructure, which is typical for the species.
- It should be emphasized that when using EM of clinical and pathological materials, the bioassay for the isolation of the particularly dangerous agent C. burnetii can be avoided, which is an operation that carries a risk of the contamination of the laboratory personnel.

In realizing our experimental infection of laboratory animals with field materials or Coxiella strains, the EM studies of organs and tissues on two methods are effective and provide valuable information about the localization of the agent, the multiplicity of the infection, and the pathological lesions. The visualization of the agent as a large cytoplasmic inclusion in the brain and spleen of white mice and guinea pigs is a convincing demonstration of the EM techniques. Our research shows that EM is very useful for the display and identification of isolates of C. burnetii in CE and cell cultures both because of the clear visualization of the agent and because of the opportunity for the early demonstration of the agent, when it is still at an initial infection state.

Our studies were the first application and description of the EM method for diagnosing coxiellas by negative staining. The fast EM method for the diagnosis by ultrathin sections is effective for the direct indication and identification of C. burnetii and its inclusions in the clinical, pathological, and experimental diagnostic materials. EM express diagnosis of Q fever shows a great performance when unconcentrated samples suspensions are prepared for 5–10 min and concentrated for about 2 h. The negatively stained preparations to conduct microscopy are prepared for 2–3 min, and the observation of a preparation in EM also takes about 2–3 min. The express EM method is extremely fast and leads to a shortening of the deadlines for diagnosis. In comparison with other diagnostic methods, the ultrathin sections method also leads to a reduction of the time for diagnosis.

Both EM methods completely eliminate the artifacts that abound the LM preparations. The number of positive results is increased sharply. The effect of EM-developed diagnostic methodshas a diagnostic, economic, epizootiological, epidemiological, and preventive impact.

The economic effect is determined by two main advantages: (1) faster and more accurate diagnosis, which is essential for organizing measures for the control of the disease and for the prevention of endangered animals and humans; and (2) sharply reduces the costs of diagnostic culture by infecting CE, passages of coxiellas, and bioassay. For an hour of EM, one is able to investigate 10–15 preparations of various animals. The high efficiency of the electron methods has necessitated their widespread implementation in veterinary diagnostic and epidemiological practice and in experimental studies.

5

Isolation and Cultivation
of Coxiella burnetii

Cox and Bell [150] found first that the pathogen multiply the yolk sacs of developing chicken embryos. Today, CE are a very suitable model for the isolation and the cultivation of C. burnetii [409]. All strains can be grown in 5–7-day-old CE. Inoculated eggs are incubated at 37°C. Depending on the concentration of the agent in inocula, CE die 2 to 14–15 days after injection, normally fallen to 4 to [41, 44] 5 days [409] embryos are not taken into account. Chicken embryos develop very typical pathological changes as a result of the infection [44, 125, 597]. In infected chicken embryos, hyperemia and stuffing have been observed. The yolk is of a thin consistency. Embryos show edatrophy and hemorrhages in the head and limbs. In some cases, the accumulation of sufficient quantities of the agent needed several passages. The sensitivity of CE to an infection with C. burnetii, especially in the starting materials with a low multiplicity of infection, is increased by a prior exposure to X-rays [34, 139]. Alexandrov et al. [72] infected CE immunosuppressed by gamma radiation. Riley et al. [436] injected CE into the yolk sac colchicine immediately before inoculation of the infectious material.

Other contemporary authors successfully used the chick embryos for the isolation and cultivation of C. burnetii of the different types and nosological origin: Moore et al. [361]—abortions and stillbirths in goats in the US; Plavšić et al. [402, 403]—miscarriages and premature births in sheep and cows in Serbia; Ho et al. [244]—reproductive disorders and mastitis in cows, coxiellos is in ticks, and atypical pneumonia in humans in Japan; and To et al. [522]—coxiellos is in poultry and wild birds in Japan and others.

The isolation and cultivation of C. burnetii in YS of CE were done in specialized laboratories by an experienced personnel in accordance with the rules of biosecurity.

The same goes for cell cultures (CC) and laboratory animals. The method of infecting CE in YS gives possibilities of obtaining large amounts of infectious material necessary for research, diagnostic, and industrial purposes (antigens, antisera, diagnostic kits).

The morphological identification of C. burnetii isolates in CE at this stage is at the level of light microscopy [409].

The reported cultivation of some viruses and chlamydia infection of CE in a chorioallantoic cavity is not developed with C. burnetii. It can be expected that this method is promising in order to obtain large amounts of the agent, which may conveniently be purified.

According to Plavšić [402], C. burnetii cultivate well in embryos of pheasants.

Guinea pigs have long been used as an experimental biological model in the diagnosis and studies of Q fever. In the experiments, adult animals weighing from 350 to 650 g were used. Infections were carried out mainly by the intraperitoneal route. Methods of subcutaneous, intranasal, and testicular inoculation [43, 244, 286–288] may also be applied. Suspensions from different source materials or passage may also be used. A permanent clinical indicator for the infected guinea pigs is an increase in internal body temperature ($\geq 40°C$) from 6 to 14 days after the inoculation. The clinical evolution of the infection depends on the dose and the concentration of the pathogen in the inoculum. According to Maurin and Raoult [343], using Coxiella suspensions containing $<10^6$ infectious units leads to the complete recovery of the guinea pigs from the disease in the course of 2–3 weeks. Bacteraemia is set for 5–7 days. A frequent pathological finding is splenomegaly. The spleen contains coxiellas, without difficulty cultured after inoculating the yolk sacs of CE. There is a specific serological response. Although acute Q fever in guinea pigs is common, a latent infection of these animals can be observed after the inoculation of suspensions with a low titer of C. burnetii. For experimental purposes, some authors have isolated and cultivated the agent in immunosuppressed guinea pigs after treatment with cortisone [475], X-ray irradiation of the whole body [476] or cyclophosphamide [82].

Not only the clinical course, but also the nature of pathological changes are dose-dependent of C. burnetii and the method of inoculation [286]. The intraperitoneal infection often leads to granuloma formation in the liver. This pathological finding in guinea pigs corresponds to the granulomatous hepatitis in humans affected by Q fever. In intra-nasally infected guinea pigs, prevalent is an inflammatory cell that infiltrates the lungs. The histopathological picture of myocarditis was observed only in animals infected with the higher dose of the pathogen in the range of $\geq 10^5$ IU [286].

From the data, it can be shown that guinea pigs are a suitable model for the cultivation and diagnostic studies of acute Q fever in animals and humans. These are particularly suitable for the primary isolation of C. burnetii from highly contaminated clinical and pathological materials, such as placentas or bodily fluids with low concentration of the agent [409]. Guinea pigs were used as a biological model of chronic Q fever in humans (endocarditis) as the disease is induced in animals, heart valves which have been previously damaged by electrocoagulation [287].

It is believed that the white mice are significantly more resistant to infection with C. burnetii, than the guinea pigs [47, 288, 409]. This concept is based on the frequent cases of asymptomatic infection, including the absence of hyperthermia. Intraperitoneal inoculation of the pathogen as a rule leads to an infection of mice, accompanied by inflammatory lesions in the spleen, liver, kidney, and adrenal glands. The spleen is enlarged several times when C. burnetii is found. Spleen extracts agent was successfully cultivated in YS of CE. The second commonly used method is the intranasal inoculation of mice, in which, although an infection was realized, some of them remain asymptomatic. The remaining mice developed a visible infection. The spleen is enlarged several times and coxiellas are found with spleen extracts agent successfully cultivated in YS of CE. The remaining mice apparently developed respiratory signs. An autopsy revealed pneumonia. Other ways of infecting white mice—cerebral, intravenous, and subcutaneous—are less explored. In an earlier publication, Kechkeeva and Kokorin [33] reported good results with intravenous inoculation and found that the sensitivity of the animals to the infection is not influenced by age, in contrast to the intraperitoneal route. The resistance of mice to C. burnetii infection decreases when applying impacts of a different nature. Alexandrov et al. [72] achieved immunosuppression in mice by gamma radiation and the successful isolation of the strains in such animals. The resistance of mice to C. burnetii infection reduces the use of impacts of different nature. Nagaoka et al. [373] achieved the isolation of the pathogen from the vaginal secretions of cats after intraperitoneal inoculation of SPF mice treated with cyclophosphamide and killed three weeks later. Atzopien et al. [82] used cyclophosphamide in suppressing the immune system of the mice prior to their challenge with strain Nine Mile.

Attempts at the isolation of C. burnetii with varying degrees of success have been done in other animal species—ground squirrels, hamsters, rats, rabbits, and monkeys. In earlier experiments with hamsters, intraperitoneal inoculation was performed, subcutaneous and intradermal route, and after the inside-testes inoculation [56]. In earlier experiments, sousliks were infected by intraperitoneal, subcutaneous, and intradermal route and after the inside-testes inoculation [56]. According to Stoenner and Lackman [500], hamsters are suitable for isolating C. burnetii that accumulate in the spleen. Monkeys are susceptible to infection by an aerosol way, but this method is not widely used in laboratory culture practice [210].

Coxiella burnetii successfully infect different cell cultures (CC) when cell lines and primary CC are used. The indication of the multiplication of the pathogen was demonstrated by means of light microscopy of preparations from the cell monolayer stained according to Ginenez or by immunofluorescent method. Sensitive to C. burnetii are lines derived from fibroblasts-mouse (L-929); chicken; human embryonic lung cells (HEL); murine macrophage-like cells–lines J774, P388D1, tumor cells, verocells, and line OL monkey kidney [70, 86, 88, 218, 280, 340, 444]. Other cell cultures are also used: HELA, HEp-2, 6-Detroit, WI-38, DBS-FRhl, and others [384, 403]. Willems et al. [553] and Ho et al. [244] isolated the agent in cell culture BGM (buffalo green monkey) after infection with clinical material from cattle, dogs, cats, ticks, and humans. Some authors isolated C. burnetii in rare cell lines derived from tissues of mosquito (Aëdes albopictus) and moths (Anthereae eucalypti) and in cell culture tissue of the tick Hyalomma asiaticum asiaticum [430, 575]. Some authors used a rare CC infection coxiellas: cell lines derived from the tissues of mosquito (Aëdes albopictus) and moths (Anthereae eucalypti) and cell culture tissue tick Hyalomma asiaticum asiaticum [430, 575].

In cell cultures, there is an issue of increasing the sensitivity to contamination by using various methods—infection by centrifugation or an improvement of the growth media [207, 369, 418, 486] and others.

5.1 Strains Isolated from Sheep

5.1.1 Abortions, Stillbirths, and Births to Non-viable Offspring

In the study of clinical-pathological materials from 107 sheep (aborted placentas, cotyledons, parenchymal organs from fetuses) originating from 71 villages in 12 regions, 20 isolates C. burnetii were received [595]. Positive culture results in chicken embryos were found in 10 settlements in 7 areas. In 8 of the cases, the isolation was made from flocks where the infection had been proven by serology, in 3 by immunofluorescence, in 7 by light microscopy, and in the other 8 cases by direct electron microscopy on the same placentas and fetal parenchymal organs. Thus, the C. burnetii nature of abortion and other manifestations of abnormal pregnancy were convincingly demonstrated by the complex use of cultural, morphological, and serological methods.

Isolation was achieved in 6–7-day-old chicken embryos by inoculation into the yolk sac of 10% suspensions prepared from cotyledons, placental tissue, or organs from fetuses liver, spleen, lung, pericardium.

Chicken embryos infected with the Coxiella isolates that died specifically between the 5th and 13th day after the inoculation showed pronounced atrophy, hyperemia, cyanosis, and petechial hemorrhages in the head and limbs. Yolk sacs of CE were sealed and hemorrhagic and their yolk mass separated without difficulty. By LM in preparations stained by Stamp or Macchiavello were discovered coxiellas with rod or oval shape measuring 300–400 nm and red or violet color. The EM in ultrathin sections showed many Coxiella cells with typical ultrastructural peculiarities.

A number of strains showed pronounced pathogenicity of CE still in its primary infection (Table 5.1). An important methodological point in the isolation and cultivation of C. burnetii was the correct selection of the starting materials, freshness, and gross pathological changes. The infectious titer of the strains was in the range of 10^{-5}–10^{-6} ID_{50}/ml.

Our experiments indicate that the C. burnetii isolates in YS of CE do not show a natural adaptive ability to cultivation in the allantoic cavity of chicken embryos. That finding is valid also for the direct inoculation in the allantoic cavity of starting suspensions of placentas and fetal parenchymal organs in attempting primary isolation of the agent. A further passage in allantoic fluid results in the presence of a single coxiellas, but sufficient accumulation of the agent was not found.

Isolated YS of CE strains of C. burnetii were identified morphologically, based on typical staining properties, the light microscopy finding, and the typical ultrastructure of the agent observed in the electron microscope. The identification of the agent was based also on its pathogenic properties for CE, guinea pigs, and white mice and serological evidence of antigen of C. burnetii in CFT with known positive sera.

Coxiella burnetii isolated directly in seronegative for Q fever guinea pigs by intraperitone alinoculation of 10% suspensions prepared from placentas, cotyledons, and organs in dose of 4–5 ml. The clinical and pathological changes are described in separate sections. The spleens of guinea pigs killed in the febrile stage—usually 4–12 days after infection showed in EM rich accumulation of the agent [595]. The presence of the pathogen was already observed by light microscopy of impression preparations of spleen and from the gray-whitish deposits on the surface of the same and other parenchymal organs, as well as the peritoneal exudate.

Table 5.1 Isolation and cultivation of Coxiella burnetii in chicken embryos in cases of sheep abortions, stillbirths, and non-viable lambs

Isolate	Passages	Inoculated CE (No.)	Dead CE (No.)	Days and Daily Lethality of CE												
				1–3	4	5	6	7	8	9	10	11	12	13	14	15
B	0	25	20	0	1	1	1	1	3	4	5	2	1	1	0	0
(Sf)	I	25	22	0	0	1	2	2	5	5	4	1	1	1	0	0
	II	25	25	0	0	0	3	4	6	6	3	2	1	0	0	0
Ch	0	20	19	1	1	2	2	3	2	1	1	1	1	2	2	0
(Vr)	I	20	19	0	1	1	4	2	1	2	2	2	2	1	1	0
	II	20	20	0	0	0	3	3	2	3	3	3	2	1	0	0
D	0	25	21	2	1	1	3	4	4	4	1	1	0	0	0	0
(Dch)	I	25	23	1	2	1	3	4	5	5	1	1	0	0	0	0
	II	25	25	0	1	1	5	5	5	6	4	2	1	0	0	0

Except for the primary isolation of the agent from placentas and fetuses, guinea pigs have proven to be a suitable biological model. Intraperitoneal infection with strains already propagated in chicken embryos or passage material (a suspension of the spleen) from the previously infected other guinea pigs. As a result, a generalized infection with C. burnetii coupled with positive serological response, histopathological changes in the spleen and liver, and accumulation of the agent in them were occurred.

The results of our experiments on the isolation and culture of C. burnetii in guinea pigs after a subcutaneous infection show that this method is also applicable. We watched the development of the disease with a longer incubation period and more scarce clinical picture and pathogen detected insufficient concentrations mainly in the spleen. Using suspensions from infected spleens, we made passage of the agent again in guinea pigs or in chicken embryos.

Inoculation of young white mice weighing 7–8 grams was performed under ether anesthesia by intracerebral or intraperitoneal route. There were injected 10% suspensions of aborted placentas and fetal organs or infected YS of CE. The control over the effect of the inoculation was a two-step one: a preliminary, whereby electron microscopy evidence of C. burnetii in significant concentrations in the suspensions could be shown, and secondary— after the injection of mice, performed in different periods depending on the route of infection and demonstrated clinical signs.

In the intracerebral challenge with both types of suspensions, the vast majority of the mice died between days 4 and 8 days after inoculation. By electron microscopy, in the brain C. burnetii was found in the form of massive intracellular inclusions consisting of a large number of cells (see Chapter 4). These results indicate that intracerebral inoculation of mice is a reliable indicator of the pathogenicity of strains investigated and sensitive method for the isolation of C. burnetii from sheep with pathology of pregnancy. In intracerebral infection with the two types of suspensions, the vast majority of the mice died between the 4th and 8th day after inoculation in the brain, revealing C. burnetii in the form of massive intracellular inclusions consisting of a large number of koksielni cells (see Chapter 4). These results indicate that intracerebral inoculation of mice and humans is a reliable indicator of the pathogenicity of strains investigated and sensitive method for the isolation of C. burnetii from sheep with pathology of pregnancy.

By intraperitoneal inoculation, the mice showed a significant resistance and poorly developed or absent clinical signs, and lethal cases were exceptional. The usual research plan included killing of animals between the 8th and 13th day, pathological view, studies of preparations of spleen and liver by CM or EM IF, and subsequent passages in white mice or CE. Upon successful isolation of the pathogen, usually after 2–3 consecutive passages in mice, we observed the presence of splenomegaly and pathogen mainly in the spleen, and to a lesser extent in liver. Part of intraperitoneally inoculated mice observed clinically for 1 month, after which they were tested serologically. In blood serum were found antibodies against C. burnetii. These results lead to the conclusion that intraperitoneal inoculation of mice can be part of the diagnostic arsenal of Q fever in abortion in sheep, but generally, the method is less effective for the isolation and cultivation of the strains compared to intracerebral inoculation and other tested patterns.

In flocks of sheep with a high percentage of seropositive animals and titer levels indicating the presence of active infection, we undertook attempts to isolate C. burnetii of the very positive blood samples (serum + coagulum). As a rule, in these herds had increased the number of abortions, stillbirths, and births of puny non-viable lambs and clinically healthy pregnant animals or normal birth.

Focused attention mainly to areas known to long-term agricultural and massive agricultural foci of Q fever (Kn, Pk). Attempts to isolation in chicken embryos have done through inoculations of 10% suspensions in the form of pooled blood samples of 5–10 seropositive sheep. Infected chick embryos showed lethality in the usual C. burnetii model—between 6 and 14 days, and in the preparations of the yolk sac was observed agents with its characteristic morphology. The isolates were identified as C. burnetii according to the criteria described above. In this experimental and unconventional way achieved isolation of the agent in 12 cases as received positive culture results represent actual herd diagnosis of Q fever in the respective holdings [597].

Attempts to cultivation of C. burnetii in cell cultures were conducted with laboratory strain "Chilnov" repeatedly passaged in YS of CE and with our isolate B (Sofia), also passaged in CE. Inoculating a 48-hour primary cell culture of chicken embryonal fibroblasts/CEF with two types of suspensions: 10% native and concentrated and purified by differential centrifugation. Removal of the individual glass panes with a cell mono layer performed on 6th, 12th, 24th, 30th, 36th, 48th, and 72th hour after the inoculation.

Using native suspensions did not lead to a cytopathic effect. Electron microscope studies of ultrathin sections of monolayer showed that up to 36 hr C. burnetii cells could not be observed. At 48 and 72 hours, they were seen as spherical and rod-like coxiellas in too small concentrations located in the cytoplasm of small groups or individually. Therefore, in the method of infecting a cell culture of primary CEF with suspensions of strains isolated from aborted sheep, it was realized a slight infection. Notwithstanding the limited accumulation of the agent, the collected cell culture fluid at 48 and 72 hours induce a specific mortality of chicken embryos, and the presence of the agent in them was confirmed microscopically.

In the second part of the experiment—inoculation of the same primary cell culture with concentrated and purified suspensions, the results were different. In the cell monolayer on the 36th and especially at 48 and 72 hours, we observed a large accumulation of Coxiella cells located intracellularly as large, medium–large, and small groups often surrounded by cytoplasmic vacuoles. Some cells were damaged and torn, and the agent was present extracellularly. This result leads to the conclusion that the infection of primary cell cultures CEF with concentrated and purified suspensions of C. burnetii from abortion in sheep resulted in a marked infection with a large accumulation of the agent and cytopathic changes in the cell.

Based on the above observations, we investigated the possibility of direct isolation in primary cell culture CEF of C. burnetii of field materials—aborted sheep placenta, cotyledons, and fetal parenchymal organs. Prior suspensions were controlled by EM for the presence of coxiellas, wherein it is found that the highest concentration of the agent is in the cotyledons, followed by the placenta, and in the spleen and liver of the fetus a medium or weak accumulation of the agent was found.

The results of inoculation of native suspensions confirmed the ineffectiveness of the method, especially in experiments on parenchymal organs, wherein the multiplicity of infection was low.

Infecting the cell culture with concentrated and purified suspensions of the same materials had a positive result. The sensitivity of the method was highest at inoculate suspensions of cotyledons, the second from placentas, and the third-by parenchymal organs of fetuses, which required additional passages for amplification of the agent.

Conducted cultural studies found mixed infection–Q fever and Chlamydia on a farm. Attempts to isolation of C. burnetii were negative in 86 sheep from 61 villages; in 44 of them were diagnosed infection with other important causes of abortions, Chlamydia abortus, and 42 settlements were free of Q fever and Chlamydia-induced diseases [595, 597].

5.1.2 Pneumonia in Sheep

Seven isolates of C. burnetii were received from inflamed lungs: 3 from lambs aged 1–6 months, 2 from older lambs (12–14 months), and 2 from adult sheep. In one of the sheep simultaneously isolated C. burnetii from blood inoculated intracerebrally in newborn mice. These were animals with severe pneumonia, which ended in death or slaughter for the purpose of diagnosis. In most cases, pathological material originates from farms with serological evidence of infection Q fever [601].

Isolates were grown well in chicken embryos as direct primary isolation was possible in the presence of a higher concentration of the pathogen in starting suspensions of pneumonic lung sections with or without the addition of parts of the mediastinal lymph nodes. In some cases, it has been required successive passages from 1 to 3 times.

The newborn mice inoculated with blood showed high sensitivity and ended fatal. In the brain were discovered large compact inclusions of the causative agent (see Chapter 4).

Direct intraperitoneal inoculation of guinea pigs with suspensions of lungs were not performed. However, inoculation of suspensions from infected yolk sac caused disease in laboratory animals, preselected as seronegative for Q fever and having a normal body temperature three days preceding infection. Guinea pigs became febrile (over 40°C) and responded with the formation of CF-antibodies against antigens of C. burnetii.

The isolates were identified as C. burnetii based on their morphological, cultural, and antigenic properties.

5.1.3 Mastitis in Sheep

In sheep with acute mastitis as a distinct and self ongoing clinical form of Q fever isolated three strains of C. burnetii, which is described for the first time in the literature [597, 602]. Isolations were made in YS of CE after inoculation of milk secretions, strongly modified by the inflammatory process.

Attempts to isolate the agent were targeted if clear evidence of widespread and active C. burnetii infection in herds, previously performed directly EM indication of coxiellas in mammary secretions and the absence of contagious agalactia and other bacterial infections.

The process of isolating and culturing the infectious agent is presented in Table 5.2. This indicates that the three strains show a similar pattern of lethality of CE. Along with the advancement of passages reduces so-called non-specific mortality of embryos in the first 4 days after inoculation. We observed a gradual increase in the amount of the specific mortality between the 6th and 14th day with a maximum of this index between 7th and 11th day. Infectious titer of strains of CE was in the range of 10^{-6}–10^{-7} ID_{50}/ml.

Table 5.2   Isolation and cultivation of Coxiella burnetii upon mastitis in sheep

Isolate	Passages	Inoculated CE (No.)	Dead CE (No.)	Days and Daily Lethality of CE												
				1–3	4	5	6	7	8	9	10	11	12	13	14	15
Tutrakan–1	0	20	18	4	1	1	1	2	2	2	1	1	1	1	1	0
	I	20	18	2	1	1	2	2	3	2	1	1	1	2	0	0
	II	20	19	1	1	0	2	3	4	3	2	2	1	0	0	0
	III	20	20	0	0	1	2	4	4	4	3	2	0	0	0	0
Târnovtsi (Ss)	0	20	17	3	2	1	1	1	1	2	1	2	1	1	1	0
	I	20	18	2	1	1	1	1	2	2	2	2	2	2	0	0
	II	20	20	1	0	1	1	2	3	3	2	3	2	2	0	0
	III	20	20	0	0	0	2	3	3	4	3	3	1	1	0	0
Dragalevci (Sf)	0	20	16	3	2	1	0	1	1	2	1	1	1	1	1	1
	I	20	17	2	1	1	1	1	2	2	2	2	1	1	1	0
	II	20	18	1	1	1	1	2	2	3	2	3	1	1	0	0
	III	20	20	0	0	1	2	2	2	2	3	3	4	1	0	0

Further study of C. burnetii isolates from cases of mastitis included traditional models of experimental infection—guinea pigs weighing 350–400 g and white mice (7–8 g). In guinea pigs, the intraperitoneal inoculation of infectious starting milky suspension lead to infection of the animals, accompanied by fever and changes in general condition as well as to pathological lesion in parenchymal organs and the production of antibodies against C. burnetii. Similar results were obtained when inoculated with infected yolk suspensions, particularly after pre-concentrated and purified by differential centrifugation. Therefore, C. burnetii isolated upon mastitis in sheep showed significant pathogenicity for guinea pigs and expressed adaptive ability for cultivation in this form after intraperitoneal infection.

Experiments with young white mice included three schemes inoculations with infectious yolk suspensions—intraperitoneal, nasal, and intracerebral.

Inoculation by intraperitoneal route leading to infection in the vast majority of cases, which was accompanied by a slight deterioration of general condition, and at autopsy between 12th and 20th day after infection—spleen enlargement in the preparations of which occurred coxiellas. Suspensions from the spleen of infected mice successfully infect CE in YS.

The nasal inoculation of mice originally applied less frequently also leads to contamination, which is asymptomatic using coxiellas not concentrated or is manifested by more frequent and obstruction of breathing by administration of the agent in concentrated form. After autopsy between 13th and 18th day found that the pathogen is localized mainly in the inflamed areas of the lungs, but also in the spleen. Note that inoculation of suspensions prepared from tissues of parenchymal organs mentioned in chick embryos have resulted in the recultivation of the agent in this biological model

Intracerebral inoculation of mice caused significant mortality between 3rd and 9th day. In the brains we observed koksieli predominantly cytoplasmatically available, but single cells are found outside the cell.

With strain Tarnovtsi (Ss) in the form of concentrated and purified yolk suspensions inoculate 48-hour cultures of BHK-21 line. At 48 and 72 hours in the cell monolayer was discovered satisfactory accumulation of the agent. Most of coxiellas were located in the cytoplasm and single cell outside it. Besides the presence of C. burnetii in infected cells were detected degenerative and destructive changes. Similar findings were made after infection of primary cell culture of chicken embryonic fibroblasts [597, 602].

5.1.4 Latent Coxiellosis in Sheep

Faeces of latently infected asymptomatic sheep in active outbreak of Q fever used for intraperitoneal inoculation of guinea pigs. The isolated strain of C. burnetii adapts easily to YS of CE following inoculation of splenic suspension of guinea pigs.

5.2 Strains Isolated from Goats

5.2.1 Abortions, Stillbirths, and Births to Non-viable Offspring

In three regions (Pz, Lch, and Sf) of outbreaks of Q fever in goats revealed by serology were isolated three strains of C. burnetii. Two of them were from placentas and stillbirth and one of parenchymal internal organs of kids, born non-viable and died the next two days [594]. The strains were identified more at the zero passage and cultivated very well in YS of CE. The same starting suspensions inoculated in the allantoic cavity of CE multiplied too sluggishly at zero and subsequent serial passages. Expectations that isolate in YS in the allantoic cavity cultivation will adapt better and did not materialize. In problem herds were not found bacterial or chlamydial participation in the etiology of these clinical conditions.

In other two flocks of goats from two regions (Kn, Sf) without serological evidence of Q fever, our attempts to isolate the agent were fruitless. In one of them by direct EM of negatively stained preparations from aborted placentas was proved chlamydial infection.

5.2.2 Pneumonia in Goats

From pneumonic lungs of goats were isolated two strains of C. burnetii. Again, insulations were in the yolk sac of chicken embryos. The materials were from animals in herds with serologically proven Q fever and mass manifestation of respiratory symptoms [594].

5.2.3 Mastitis in Goats

Inflammation of the mammary gland of goats, especially subacute and oligo-symptomatic forms, are often seen as part of a general clinical syndrome of Q fever, and the release of the pathogen with secretions and excretions is the subject of etiological and epidemiological studies.

We turned our attention to the acute mastitis as dominant or sole symptom indicating clearly prominent clinical manifestation. Such cases after pre-liminary serological identification of Q fever in the herd, especially in the individual animal, identified as very suitable for the isolation of the causative agent of milky discharge and its characterization [594, 597].

Coxiella burnetii PM strain was isolated from goat milk with acute mastitis and studied by immunological, biological, and serological methods. The properties of this strain was compared with those of the reference strain Henzerling, isolated from man and our two strains of cattle—HP isolated from the placenta of aborted cow and MP from placenta of a cow with normal birth.

Immunological identification of goat strain PM was done and comparison with strains HP, MP, and Henzerling was conducted in guinea pigs. The animals in each formed four groups were infected with one of the above four strains, and one month later performed a new infection, but separately with each of the other three strains [594, 597, 631]. The starting material for the first intraperitoneal inoculation at a dose of 2 ml represented a 10% suspension of spleen, liver, and lungs of guinea pigs infected with the respective strains in consecutive passages in these laboratory animals as follows: PM and HP—6th; MP—7th; Henzerling—4th (Table 5.3). This shows that none of those were infected with PM, HP, and Henzerling guinea pigs after reinfection on the 30th day with the other strains showed no signs of disease. This speaks for full mutual cross-immunity between studied goat strain both bovine strains

Table 5.3 Results of immunoassay of guinea pigs with strain PM isolated from goat affected by Coxiella burnetii mastitis, bovine strains HP and MP, and reference strain Henzerling

| First Infection Strain | No. Guinea Pigs | Second Infection | | | | | | | | | | | |
| | | HP | | | MP | | | PM | | | Henzerling | | |
		No. Guinea Pigs	Total Days of Fever or Other Clinical Signs	% Days of Fever	No. Guinea Pigs	Total Days of Fever or Other Clinical Signs	% Days of Fever	No. Guinea Pigs	Total Days of Fever or Other Clinical Signs	% Days of Fever	No. Guinea Pigs	Total Days of Fever or Other Clinical Signs	% Days of Fever
HP	–	4	–		4	0		4	0		4	0	
MP	4	–	2	0.5	–	–		4	0		4	0	
PM	4	4	0		4	0		–	–		4	0	
Henzerling	4	4	0		4	0		4	0		–	–	
Uninfected control group	4	4	26	6.5	4	21	5.2	4	18	4.5	4	17	4.2

and strain Henzerling. The exception made under strain MP after reinfection of the guinea pigs with HP, which registered a fever for two days (Table 5.4).

As a control group of 16 uninfected guinea pigs, which in the same period (30 days) inoculated with suspensions of strains PM, HP, MP, and Henzerling. Lack of immunity in the control animals leads to infection with the four strains, accompanied by fever and other clinical signs. It should be noted that strain PM events pathogenicity for guinea pigs with most indicators similar to those of the reference strain Henzerling (Table 5.3).

Serological results are presented in Table 5.4. We used CFT as each antigen checked with antisera of other strains. Antigens received in YS of CE infected with strain PM (16th passage), HP (7th passage), and MP (11th passage). Specific immune sera from guinea pigs received 30 days after infection with the 10% suspensions of the strains: PM (10th passage), HP, and MP (8th passage). From Table 5.4 can be seen that are established three types of reactions, namely vertical (PM), horizontal (HP), and intermediate (MP). The vertical type is characterized by high serum titers and low antigen titers, the horizontal type has high antigen titers and low serum titers, and the intermediate type has medium antigen titers and medium serum titers [631].

The data in Table 5.5 show that the antigens of coat strain PM and the bovine strains HP and MP specifically react against antiserum to the reference strain Henzerling and back, antigen of Henzerling, specifically reacts against antisera of PM, HP, and MP. Relatively lowest CF titers (1:40–1:160) were found in the testing of antisera to antigen prepared from strain HP, medium–high (1:60–1:320) with antigen from strain MP, and high (to 1:640) with antigen from PM. In all cases, maximum titers (1:640 against PM, 1:1280 against MP, and 1:2560 against HP) were found in reactions involving antigen strain Henzerling and antisera of those strains [631].

Virulence of the strains tested in guinea pigs and white mice [597]. Guinea pigs inoculated testicular with 1 ml, and intraperitoneal and subcutaneous with 2 ml of 10% suspensions of YS infected with strain PM (17th passage), HP (8th passage), MP (12th passage), and Henzerling (unknown number of egg passages). White mice were infected by intraperitoneal and intranasal route with 0.5 ml of the strains: PM (17 passages), HP (six), MP (thirteen), and Henzerling (numerous passages unknown number). Infectious titers of the strains were 10^6 ID_{50}/ml.

The results of experiments using guinea pigs showed that the strain isolated from the milk of a goat affected by mastitis has the low virulence compared to the two bovine strains and Henzerling (Table 5.6). The caprine strain causes disease in guinea pigs after an incubation period of 16.2 days, the average height of the temperature reaction- 39.6°C lasting 3.2 days and no lethal case.

Table 5.4   Titration of antigen C. burnetii with goat and bovine origin against homologous sera from guinea pigs

Dilutions of Serum	Strain PM Isolated from Goat with Mastitis					Strain MP Isolated from Cow with Normal Birth					Strain HP Isolated from Aborted Cow				
	1:2	1:4	1:8	1:16	1:32	1:2	1:4	1:8	1:16	1:32	1:2	1:4	1:8	1:16	1:32
1:10	4	4	3	–	–	4	4	4	2	–	4	4	4	4	3
1:20	4	4	2	–	–	4	4	3	2	–	4	4	4	4	2
1:40	4	4	3	–	–	4	4	3	1	–	2	2	–	1	–
1:80	4	4	1	–	–	4	2	3	–	–	–	–	–	–	–
1:160	4	4	2	–	–	2	2	–	–	–	–	–	–	–	–
1:320	4	3	1	–	–	1	–	–	–	–	–	–	–	–	–
1:640	4	3	1	–	–	–	–	–	–	–	–	–	–	–	–

Table 5.5 Survey results of homologous and heterologous sera and antigens of strains HP, MP, and PM compared with strain Henzerling

| | Sera from Guinea Pigs at 30 Days of Inoculation | | | |
| | Titer of Complement-Fixing Antibodies | | | |
Antigens	HP	MP	PM	Henzerling
HP	1:40	1:80	1:160	1:80
MP	1:320	1:160	1:320	1:160
PM	1:640	1:320	1:160	1:80
Henzerling	1:2560	1:1280	1:640	1:320

Upon infection with a strain MP guinea pigs have a relatively shorter incubation period (13.1 days), a slightly higher average temperature (39.7°C) lasting 3.5 days, but the mortality is 50%. The most virulent strain proved HP (C. burnetii in cows), with the shortest incubation period (6.1 days), higher body temperature (40.1°C) lasting 4.6 days and the highest mortality (83.3%). The comparison of the foregoing data with established experience in virulence of Henzerling shows that the reference strain produces the highest temperature rise (40.5°C), has an incubation period of 8.3 days, and cause mortality, which is greater than that of the caprine and less compared to the two bovine strains. These results are complemented by other clinical and pathological phenomena—the result of infection with each of the C. burnetii strains studied: temperature-scrotal syndrome after testicular inoculation; inflammatory infiltrate at the site of subcutaneous inoculation; and two to three times splenomegaly after intraperitoneal inoculation.

When white mice appliqué of infectious material from caprine and bovine strains by intraperitoneal or intranasal routes led to a benign disease course with little or no visible signs and no deaths. Strain HP showed the greatest virulence as evidenced, on the one hand, the pathological picture of the 20th day after infection and finding plenty coxiellas in impression preparations from spleens, livers, and sometimes in the lungs of another. Infection with strains MP and PM led to a small increase in the spleens of the mice and the relatively small amount coxiellas in microscopic preparations of spleens and livers, but the agent was not found in the lungs. Lowest virulence for mice had a human strain Henzerling.

The results of the comparative studies show that PM strain isolated from goat affected by mastitis and both bovine strains—identified as C. burnetii— show some differences in the serological and biological terms, which is probably due to the uneven phase [597].

Table 5.6 Virulence of Coxiella burnetii strains for guinea pigs

Strain C. burnetii	Infected Guinea Pigs (No.)	Average Incubation Period (Days)	Temperature Reaction		Mortality in Guinea Pigs (No./%)
			Average Duration of the Temperature Rise (Days)	Average Height in Body Temperature (°C)	
PM—mastitis in goats	4	16.2	3.2	39.6	0/0
MP—isolated from the placenta of a cow with a normal birth	4	13.1	3.5	39.7	2/50
HP—isolated from the placenta of aborted cow	6	6.1	4.6	40.1	5/83.3
Henzerling-reference	4	8.3	6.3	40.5	1/25

5.2.4 Isolation of C. burnetii from Blood in Goat Herds with Active Q Fever

This method was applied in goat herds goats with a significant range of infection and the presence of high individual and geometric mean titers of specific antibodies. They were inoculated 7-day CE with pooled suspensions of blood from 5–10 seropositive animals. From such samples, we received seven isolates in herds of three regions (Kn, Lch, and Sf).

5.2.5 Latent Coxiellosis in Goats

In active focus of Q fever in goats isolated in guinea pigs (i.p.) strains of C. burnetii from fecal samples of goat have latent in apparent form. In the spleens of laboratory animals, pathogen accumulates in significant concentrations [594].

5.3 Strains Isolated from Cattle

5.3.1 Abortions, Stillbirths, and Births to Non-viable Calves

We examined a total of 96 clinical and pathological materials of 87 cows with abortion or other disorderly of pregnancy grown in 50 settlements of 10 regions. The number of isolates received 16 of 10 herds in 10 villages in 7 regions—SF, Sofiacity, Kn, Hs, Kj, Dch, and Lch. Coxiella burnetii was isolated after intra-yolk inoculations as suspensions of placental tissue— 6 cases, of cotyledons—5, parenchymal internal organs of fetuses—4, and uterine-vaginal secretions—1 case [591].

Almost all farms of origin of the culturally positive materials have serological established coxiellosis. Six of the isolates were from one of the largest agricultural foci of Q fever Piperevo (Kn) and have received along the four consecutive years, which testifies to the lasting C. burnetii persistent infection in cattle.

C. burnetii isolates showed similarity in cultural properties on models of CE, guinea pigs and white mice, but the effectiveness of isolation and the further passages largely depended on the concentration of C. burnetii in the source material. The most suitable were placentas and cotyledons, where with the help of electron microscopy and light microscopy were found the largest deposits of pleomorphic cells of the causative agent. As an example of the behavior of isolate, C. burnetii from abortions in cows under cultivation in these biological models, as well as serological its activity, immunological references, and virulence presented strain HP (see previous section "strains isolated from goats" item "Mastitis"). Attempts for the isolation and cultivation of C. burnetii in chorioallantoic cavity of CE included inoculations with a suspension of the above-mentioned pathological and clinical material as well as yolk suspensions already isolated strains. Multiplication of the agent in very low concentrations observed using inoculum of cotyledons and placentas and from some isolates in yolk sacs have a good accumulation of the agent. Overall, however, the results are very poor, indicating a lack of natural ability adaptation to the allantoic cavity cultivation.

Attempts to isolate the C. burnetii were negative by the test material of 71 cows in 40 locations. As a rule, these holdings were not found positive for Q fever animals or their number was minimal with a height of serological titers not indicated the presence of active infection.

Clinical and pathological materials positive for C. burnetii were negative for pathogenic bacteria and chlamydial agents. In negative for C. burnetii fetuses and placentas in 10 cases was diagnosed infection with Chlamydia, which have important significance in differential diagnosis at abortion and related clinical conditions.

5.3.2 Pneumonia in Calves

In cases of pneumonias in calves were isolated two strains of C. burnetii in chicken embryos after inoculation of suspensions from inflamed lung tissue of calves (2–4 months) from two farms [591]. In one case, the starting suspension was a mixture of lung and mediastinal lymph nodes. Strains identified by IF and EM at zero passage. Bacteriological controls were negative. In preliminary serological tests in the same farms we found considerable scope of the infection and titer values characteristic of the active agricultural outbreaks of Q fever. Other attempts to isolate the pathogen were not taken.

5.3.3 Mastitis in Cows

Our clinical-epidemiological and serological studies have shown that infections caused by C. burnetii are found in dairy cows with problems regarding the mastitis.

Attempts to isolate the pathogen involved animals with acute inflammation of the mammary gland, as previously selected seropositive for C. burnetii while microbiologically negative for pathogenic bacteria in the milk secretion. From such cases three isolates of coxiellas in chicken embryos after infection in the yolk sac were found. With suspension prepared of milk received from a cow were inoculated in parallels chicken embryos and intraperitoneally into guinea pigs. The latter were sensitive to the pathogen from milk and reacted with hyperthermia and specific antibody [591].

We cultivated the pathogens in YS of CE coxiellas after reaching a satisfactory accumulation (LD$_{50}$ log $5 \times 10^{4.5}/0.5$ ml) and increase their multiplicity by differential centrifugation, successfully infecting cell cultures—primary chicken embryo fibroblasts and stable cell line BHK-21.

5.3.4 Metritis and Endometritis

In acute and chronic recurrent endometritis, most often as a result of retained placenta, especially after abortion caused by C. burnetii, endometrial exudate contains coxiellas in high concentrations. The discovery of the agent was carried out directly by the methods LM, IF, and especially EM. In blood serum of affected cows were detected CF-antibodies. Bacteriological sterile suspensions of uterine exudate inoculated in an amount of 0.25 ml in YS of 7-day-old CE. Using this method, for the first time in our laboratory practice isolated the C. burnetii of three cows. The passages in CE has no difficulties with the typical model of pathogenicity and lethality of chicken embryos. CE-adapted strains in the form of concentrated and purified yolk suspensions cultured without difficulty in cell cultures CEF and BHK-21. With inoculum of the infected cell cultures introduced inside yolk sacs easily achieved re-cultivating of the pathogen in CE [597].

The infectious agent isolated from cases of severe septic metritis cow in a fatal outcome [591]. The infection of CE in YS achieved by inoculation with a suspension of tissue of the uterine wall and separately with inoculum from the spleen and liver of the cow. Strain (SM-1) was characterized by pronounced pathogenicity of CE and accumulation of C. burnetii cells still in primary isolation (Table 5.7).

Summed data that shown primary isolation and subsequent three successive passages in the CE show lethality between the 1st and 15th day after inoculation as so-called non-specific mortality (1–5 days) was 14.54%, and lethality induced by the specific pathogenic effect of C. burnetii (6–15 days) was 85.46%. Maximum number of CE were killed between 6th and 10th day (Table 5.7). SM-1 strain amplified in YS was easily adapted to cultivation in the cell line BHK-21 and primary cell culture CEF. Here, the contamination was significantly facilitated with the use of concentrated and purified suspensions, which led to a higher multiplicity of infection and accumulation of polymorphous cells of C. burnetii, as well as cytopathic effect.

Table 5.7 Isolation and cultivation of C. burnetii—strain SM-1 from the uterus of a cow with fatal septic metritis

Passage	CE Inoculated (No.)	Dead CE (No.)	Days and a Daily Mortality of Chick Embryos												
			1–3	4	5	6	7	8	9	10	11	12	13	14	15
0	30	30	3	2	2	3	3	5	3	2	1	1	2	2	1
I	30	30	2	2	1	4	3	4	4	3	2	2	2	1	0
II	25	25	1	1	1	3	2	3	3	3	3	3	2	0	0
III	25	25	0	0	1	4	3	3	3	4	3	3	1	0	0
Total	110	110	6	5	5	14	11	15	13	12	9	9	7	3	1

The isolation of C. burnetii from a fatal case of septic metritis in cow represents the first description in the country like that [591, 597].

5.3.5 Coxiella burnetii Isolation from the Blood of Cattle in Active Foci of Q Fever

Like in sheep and goats, cattle, especially cows from herds with high rates of seropositive reagents we isolated C. burnetii of blood in five cases. Matching blood samples (serum + coagulum), with high titers of specific antibodies which, in aggregate suspensions were inoculated chicken embryos into their yolk sacs.

The results contribute to the herd diagnosis of Q fever and can serve as an indicator of the activity of the infection in a herd or farm studied [597].

5.3.6 Isolation of C. burnetii from Wild Animal Species and Arthropods

In view of the integrated presentation of the issue of natural outbreaks of Q fever, data isolation of the infectious agent from wild animals and ticks is presented in the section on natural foci.

Summed results presented above show that from domestic ruminants are obtained 64 isolates of C. burnetii. These results are analyzed in Table 5.8. In addition, isolation of the agent in 21 cases from wild mammals and birds, but also from ticks, is shown. The latter are described in detail in the section on natural outbreaks of Q fever.

In the cultivation of new strains and laboratory strains performed several passages in which a large number were infected chicken embryos. Amount of the studies is significant. Their results show that the Coxiella burnetii infection is too widespread. The isolation of causative agent from domestic animals with different clinical manifestations is important and indisputable confirmation of serological evidence of infection with C. burnetii among the species. That finding is valid in terms of wild mammals, birds, and ticks.

The main method used in the isolation and cultivation of C. burnetii was infecting chicken embryo yolk sac. This method according to the regulations and standards of the Office International des Epizooties—OIE (World Organization for Animal Health) is a common and accepted as very suitable for the isolation and cultivation of all strains C. burnetii [381, 409]. When dealing with coxiellas of avian origin, the rules for isolation and identification of avian pathogens were strictly observed [68]. The obtained data are that this classical method used in a number of early and contemporary studies is effective.

Table 5.8   Isolation of Coxiella burnetii from mammals, birds, and ticks

Group	Species	Clinical Form	Isolates/No./
Domestic ruminant animals	Sheep	Abortions, stillbirths, deliveries of non-viable lambs	20
	Sheep, lambs	Pneumonia	7
	Sheep	Mastitis	3
	Sheep	Latent form	1
	Goats	Abortions, stillbirths, deliveries of non-viable kids	3
	Goats	Pneumonia	2
	Goats	Mastitis	1
	Goats	Latent form	1
	Cows	Abortions, premature births, stillbirths, deliveries of non-viable calves	16
	Calves	Pneumonia	2
	Cows	Mastitis	3
	Cows	Endometritis and metritis	4
	Cows	Latent form	1
Wild mammals	Wild rabbit (*Lepus europaeus*)	Coxiellosis	1
	Wild murine rodent (*Dryomys nitedula*)	Coxiellosis	1
Wild birds	Doves	Coxiellosis	2
	Pigeons	Coxiellosis	3
	Pheasants	Coxiellosis	2
	Crows	Coxiellosis	2
	Ravens	Coxiellosis	1
Ixodic ticks	D. marginatus	Coxiellosis	3
	R. bursa	Coxiellosis	2
	R. sanguineus	Coxiellosis	2
	H. plumbeum	Coxiellosis	2
Total			85

In addition to diagnosis, the method of infecting CE in YS allows harvesting of the large amounts of infectious material to research and biomanufacturing purposes [17, 44, 244, 361, 409, 438, 462, 473, 562, 563].

The application of this method enabled us to isolate for the first time in our practice C. burnetii at nine clinical conditions in domestic animals: abortions, premature births, stillbirths, and non-viable calves (cattle); endometritis and metritis (cows); pneumonia (calves, adult sheep, and lambs a year old); acute mastitis (goat); and coxiellosis in five species of wild birds: crows, ravens, pigeons, doves, and pheasants.

Isolation of C. burnetii, achieved by us at coxiellosis in wild rodents Dyromys nitedulla (Doormouse) and mass acute mastitis in sheep as a distinct and self ongoing clinical form, are the first descriptions of its kind in the literature [597].

We isolated the agent of Q fever in a number of other conditions in several species of animals: sheep abortions, stillbirths, and newborn unviable; lambs with pneumonia; goats with abortions and related manifestations of the pathology of pregnancy; goats with pneumonia; cows with mastitis; cattle, sheep, and goats with latent asymptomatic form of Q fever; and hares with coxiellosis. These findings confirm the positive results of other Bulgarian researchers: Ognianov et al. [44], Genchev et al. [17], and Pandarov [47, 49]—abortion and related conditions in sheep; Ognianov [43] and Shindarov et al. [66]—abortion and lung inflammation in goats; Genchev et al. [19] and Ognianov et al. [45]—pneumonia in lambs; Genchev et al. [19] and Ganchev et al. [13]—coxiellosis in wild rabbits. In connection with our isolation of C. burnetii of wild animals and birds, Table 5.8 will note that in earlier studies by other authors in the country, the agent was isolated from hamsters (Citellus citellus) [57, 65], badger (Meles meles), and brock [470].

The first successful isolation of C. burnetii from pooled blood samples of goats having fever (the town of Peshtera) in CE (Shindarov et al. [66]) and guinea pigs (Ognianov [43]) was a good sign of the ability of these biological models using culture diagnostic approaches. In our extensive experiments with cattle, sheep, and goats from herds of convincing serological evidence of active infection, the main criteria in the selection of pooled blood samples (serum and coagululums) were the presence of a significant number of positive sero reagents in the herd and higher individual and geometric mean titers of antibodies against C. burnetii. The number of isolates received in CE in three animal species (not included in Table 5.8)—12 sheep, 7 goats, and 5 cattle—shows the effectiveness of this additional method of herd diagnosis of active Q fever in ruminants. Undoubtedly, proof of Q fever in those kinds of farm animals by isolation of the pathogen clarifies the nature of some of the above clinical diseases that pose a threat to livestock and people.

Isolation of these nine strains of C. burnetii (seven in YS of CE and two in guinea pigs) from ticks collected from cattle, sheep, goats, and dogs expands our positive cultural findings in the development of materials suspected of natural reservoirs of Q fever. These recent data confirm the existence and persistence of a number of natural foci of Q fever in Bulgaria supported by ticks and some species of wild animals and birds [20, 21, 57, 66].

In experiments on isolation and cultivation of C. burnetii adhere to the rules for precise selection of source materials in terms of gross pathological changes freshness and number of the selected tissue slices. The primary suspensions are processed from the largest possible number of internal parenchymal organs and body fluids and in subsequent passages inoculum received from several YS with lesions and morphological control on the possible presence of the agent.

This requirement is not particularly necessary when dealing with fetal organs, in which the accumulation of the agent is not so great as in placentas on abortion and related conditions and the microscopic indication of the pathogen is difficult. In this respect, our methodological principle coincides with that of other researchers [17, 361, 502]. The approach allowed the avoidance means for increasing the sensitivity of CE to infection—gamma radiation, injection of colchicine, and others. [57, 72, 215, 438]. The only materials that inoculated in advance immunosuppressed by gamma rays CE were from heart valves to people with endocarditis. In these experiments were isolated four strains of C. burnetii [597].

In several cases, we achieved isolation and identification of the agent from different backgrounds more at the source (zero) in passage in YS of CE. Such quick isolations were reported in other publications [44, 125, 409]. According to Ognianov et al. [44], the use of fresh placentas and aborted cotyledons of sheep, timely delivered for research and proven microscopic discovery of the causative, provides isolation of C. burnetii still at primary passage. The results of our large volume studies have shown that when cultured not only with a sample of aborted ovine placenta, but also with various other raw materials; such rapid isolation and identification is possible. An important methodological point in these cases is the use of electron microscopy for early indication and identification of isolates. In difference from other authors, we used this method extensively and this has allowed us to significantly accelerate and make more reliable diagnosis of C. burnetii. This is especially necessary because of the mistakes that allowed the presence of chlamydia and brucella having similar morphology in light microscopy preparations [409]. Convincing proof of coxiellas in YS still in their primary isolation when the concentration of agent usually is not large and light microscopic diagnosis is uncertain testifies to the great advantage of EM as a method for early diagnosis of the causative agent. In the identification of isolates sticking to a set of tests and indicators: (a) morphological identification of the typical intra-cytoplasmic inclusions and cells of C. burnetii; (b) proven infectivity of the agent for CE at inoculation YS; (c) proof of pathogenicity of the agent for guinea pigs and white mice by intraperitoneal inoculation; (d) serological evidence for the antigen of C. burnetii in CFT with known positive sera; (e) exclusion of other microorganisms in chicken embryos.

Unlike infecting CE in YS, the method of isolation in chorioallantoic cavity is not developed with C. burnetii. Positive aspects of the successful cultivation in allantoic cavity achieved in chlamydia and some viruses are expressed in obtaining large amounts of infectious material from allantoic fluid—raw materials, characterized by a large degree of natural purity. Our experiments showed that both coxiellas in inoculum of placentas and fetal parenchymal organs and in isolates of the agent in CE did not show adaptation properties of multiplication in this place, especially in zero passage. Subsequent passages in the allantoic fluid were accompanied by the presence of the agent in low concentrations, but there was no expressed tendency to a significant increase in its volume. Obviously, in these cases experimenting with different means to influence CE in order to facilitate the penetration of the agent and their reproduction in chorioallantoic cavity, possible differences in pathogenicity of individual strains were taken into account.

In our research, direct isolation of C. burnetii in guinea pigs following intraperitoneal inoculation was possible with clinical and pathological material from sheep (abortion and related conditions, mastitis), goats (mastitis), cattle (abortion, mastitis, latent form). Particularly suitable for this purpose are highly contaminated samples from the field, such as the placenta, as well as body liquids (milk, etc.) having a lower content of the pathogen [409]. Very positive moment in the evaluation of the method is the regular establishment of certain findings in positive for Q fever cases: pyrexia, accumulation of the agent in the spleen, and the presence of specific antibodies against C. burnetii in the blood serum. Expressive manifestations of these indicators, especially in the clinical development and pathomorphological changes, depend not only on the method of application but also the dosage and concentration of the agent in the inoculum [286, 343]. We believe that the effect of intraperitoneal infection of guinea pigs depends more on the type of strain and its natural adaptation ability for cultivation in the body of this kind of laboratory animal. A typical example of this are our isolates from mastitis in sheep showed significant pathogenicity for guinea pigs, manifested by fever, general ill health, pathological lesions in parenchymal organs, and expressed serological response. In our other comparative experiments involving cited above two bovine strains (HP-abortion; MP-latent infection), a goat (PM-mastitis), and one human (Henzerling reference), we found differences in virulence for guinea pigs as the highest was that of abortifacient bovine strain, followed by the other bovine strain, human, and goat strain. Our positive results and our evaluation of the method of intraperitoneal infection of guinea pigs are consistent with the literature on this issue [43, 244, 286–288, 409].

Direct isolation of the agent was also possible by subcutaneous inoculation of a suspension of placental and fetal material at the abortion in sheep. The enclosed route of infection lead to clinical disease with a long incubation period and slightly manifested signs and pathogen detected in moderate concentrations mainly in the spleen.

The culture work with guinea pigs also included inoculations strains of C. burnetii, propagated in YS of CE: intraperitoneal, subcutaneous, and testicular (cattle abortion or normal birth; goat mastitis, human reference strain Henzerling); only intraperitoneal (sheep with pneumonia or miscarriage, stillbirth or weak non-viable offspring). The results were positive – agents multiply in the spleen and other parenchymal organs, and received suspensions of spleen served as the best material for the passage of new infections in other guinea pigs or chicken embryos.

Conducted experiments with guinea pigs for isolation and cultivation of C. burnetii and studies of the serological activity, some biological references, virulence, and other properties of the C. burnetii strains certainly point continuing relevance and utility of this biological model for the study of Q fever in animals. Pronounced tropism of the agent to the liver and other parenchymal organs of the guinea pigs is demonstrated by the histopathological lesions, including the presence of granulomas, definitely corresponds to the processes developing in acute Q fever in humans, leading to granulomatous hepatitis and granulomatous formations in the spleen, bone marrow, and other organs [343]. So, undoubtedly, studies on the coxiellosis in guinea pigs serve as a suitable biological system for studying human disease, especially granulomatous hepatitis [245, 256, 320, 378, 478, 503] and endocarditis [287].

White mice were also used for isolation of the agent, but there is a perception that they are more resistant to infection, compared with guinea pigs [47, 288, 409]. According to some authors [409], the sensitivity of mice to the C. burnetii is approximately ten times smaller than the guinea pig. Applied are different methods to overcome the resistance of the mice—the suppression of the immune system with cyclophosphamide, gamma radiation, and the like [72, 82, 373]. Indeed, the concept of the greater resistance to white mice is based on observations in the two ways of inoculation—intraperitoneal and intranasal. Careful analysis of the data shows that in both cases is implemented infection accompanied by inflammatory lesions in parenchymal organs of the abdominal cavity (i.p.) and lung (nasal), but the visible clinical manifestation in mice is usually moderate or slight. Asymptomatic course of infection is also observed [47, 288]. Outside that biological granted for greater resistance of mice is essential inoculation source material, the concentration of the pathogen, the infectious titer, and natural adaptive ability of the strain to reproduction in the body of the mice at the chosen route of administration of the infectious material. Scarce data there is about other ways of infection—intravenous, subcutaneous, and cerebral. According to Kechkeeva and Kokorin [33], the intravenous challenge with coxiellas is successful in mice of different ages. Our experiments covered three schemes of inoculations: intracerebral, intraperitoneal, and intranasal. It was achieved direct isolation of C. burnetii in the brains of newborn white mice and young white mice after inoculations of blood from sheep and hare, and as placentas, cotyledons and parenchymal organs of fetuses during abortions in sheep. Results obtained—clinical disease with high lethality and massive intracellular accumulation of the agent in the brain—show the reliability and sensitivity of the cerebral infections and a method for isolation of C. burnetii from ovine abortion and related conditions of the pathology of pregnancy [597].

Only, but successful experiment with blood of hare also confirms the potential of the method for isolation of C. burnetii.

The direct isolation of infectious pathogen of field materials in abortion in sheep after the intraperitoneal inoculation of young white mice is also possible. Clinical signs were less developed. Single infected mice showed no visible signs of illness. Mortality was minimal. The main location of the agent found in the spleen as larger deposits registered after 2–3 ip passages in mice. Serological response was convincing. As described above, a feature of intraperitoneal method for isolating C. burnetii in white mice of field materials allows us to accept the method as intermediately susceptible. However, this method has its place in the spectrum of diagnostic maneuvers for proof of Q fever in animals. This opinion is based on the results of our efforts with isolates the agent of CE from sheep (miscarriages, stillbirths, no viable lambs, mastitis), inoculated i.p., but also in other experiments—cerebral or nasally; goats with mastitis—i.p., nasal; cattle (abortions and latent infection)—i.p., nasal; people (reference strain Henzerling)—i.p., nasal. Our observations show that inoculation of mice by the intraperitoneal and intranasal routes of all strains lead to infection and disease of benign clinical course without mortality, but lesions were observed in the parenchymal organs, and the presence of C. burnetii therein. Last served as a good material for passage in mice and recultivation in CE. Intracerebral inoculation with sheep strains grown in YS of CE, like fieldwork material from sheep, was highly effective. Therefore, cerebral infection with strains tested is a reliable indicator of their pathogenicity on a model young white mice.

Within the overall characterization of benign course of the disease by intraperitoneal and intranasal inoculation of mice, we observed differences in the comparative study of the virulence of the aforementioned two bovine strains HP and MP, caprine PM, and human Henzerling. And in experiments with white mice at high virulence showed PM isolated from aborted cow, lower MP from latently infected cow gave birth to normal, and the lowest—PM—isolate from goat's milk and Henzerling. We believe that the results of studies of the aforementioned strains by immunological, biological, and serological methods should be interpreted according to their phase state. Set different values of CF antibodies showed that C. burnetii strains were not in the same phase state; hence, their antigens induce production of "early" or "late" antibodies. Data obtained by titration of antigens and research with homologous and heterologous antigens and sera of the tested strains and the control reference strain warrant the conclusion that the strain HP, passaged 8 times in CE is in phase I while strain MP (11–13 passages), is in transitional phase and strain PM (16–17 passages) is in phase II. Therefore, higher titers of specific antibodies were ascertained with an antigen prepared from strain PM against antiserum derived from strain HP, while the lowest titers are found in the reactions with the antigen from strain HP and antisera from other strains. These results are somewhat consistent with the data of Literak [300, 301] and Jarmin [67] obtained on blood sera of cows, calves, sheep, and dogs in endemic regions Q fever. The results of biological tests correlate with those of the cross-immunological and serological reactions. Infection of guinea pigs inoculated with finding in phase I strain HP has a much more severe course, a shorter incubation period, higher rates of temperature rise, longer development, and higher mortality. In contrast, the infection induced by strain PM /phase II/ is characterized by milder symptoms, significantly longer incubation periods, less temperature increase, and no lethal cases. Trials with white mice support the view to lower virulence of C. burnetii in later passages in CE, which is associated with the transition of the strains from one phase state to another [110]. According to some authors [50, 61, 109, 222, 290, 501], phase of C. burnetii determine various properties of the pathogen: virulence, immunogenicity, serological activity, resistance to phagocytosis, and others. Enright [181] expressed the view that the species from which the agent is isolated also affects the degree of its infectivity. In his study, strains isolated from cattle were more virulent for laboratory animals than sheep strains [181]. We will emphasize that our research on immunological, serological, and virulence properties of strain C. burnetii, isolated from the milk of a goat with mastitis, is the first such description in the literature. The data about these properties of the strain showed similarity with the properties of the reference strain Henzerling. Like guinea pigs, the experience gained from the experimental infection of white mice for diagnostic and studies of C. burnetii and Q fever mainly in livestock is used for the development of animal models of chronic Q fever in humans. There are two experimental model of endocarditis—immunocompromised mice with cyclophosphamide [82] and pregnant white mice [343]. Recently, Stein et al. (2005) [496] reported a model of acute Q fever (pneumonia) which was achieved aerosol infection with strains Nine Mile and Q212, manifested by clinical signs and lung lesions.

Rabbits are rarely used for the isolation of C. burnetii. Pregnant animals showed high sensitivity to contamination, which often leads to stillbirths [343]. One of the four experimental animal model for studies of endocarditis in humans has been introduced in rabbits with a catheter in the left ventricle, leading to damage to the aortic valve and thrombotic vegetation. Subsequent i.p. inoculating agent leads to its localization at the site of the lesion and the development of endocarditis due to C. burnetii [363]. The review of the scarce literature data showed unlike guinea pigs and white mice, the minimum amount of culture working with rabbits and a lack of systematic observations do not provide a basis for more in-depth assessment of the capabilities of the method for isolating, cultivating, and studying the pathogen and disease. We experimented successfully three ways of infecting domestic rabbits—intravenous, intraperitoneal, and intra-testicular with coxiellas from direct materials (blood from sheep in bacteremia, concentrated, and purified placental suspensions of aborted cows and strains propagated in YS of embryonated eggs from cases of mastitis and abortion in sheep and cows with metritis). Significant methodological element in these experiments was scrutinized prior verification of the agent sufficient multitude in the starting inoculum. All infected animals implemented clinically with hyperthermia and fever, pregnant rabbits—abortions and the animals inoculated intra-testicularly developed orchitis and peri-orchitis. The implementation of infection confirmed directly in parenchymal organs, uterus, and placenta and indirectly through the detection of specific antibodies. The agent in the tissues of rabbits was recultivated without difficulty in CE. These results definitely show that the rabbits are suitable laboratory animals for direct isolation of the pathogen from these materials, as well as for passage of the already isolated strains and for conducting biological experiment. They give rise to the need to test the sensitivity of rabbits to infection with other strains C. burnetii having different types nosological and geographic origin, as well as clinical and epidemiological and etiological studies on Q fever in rabbits farmed.

The analysis of the literature data shows that in laboratory diagnostic work on Q fever in humans, cell culture techniques for isolation and culture of C. burnetii is used more often than in disease in animals. Standards for diagnostic tests of the World Organization for Animal Health—OIE [409] does not include a method for isolation of C. burnetii in cell cultures (CC). For research purposes, however, the method certainly has its place in the activities of the specialized laboratories. Used primary cell cultures and continuous cell lines [70, 86, 218, 340, 403, 444] and various human samples (liver biopsies, bone marrow, cerebrospinal fluid, blood, plasma, heart valves, vascular aneurysms, placenta) have been used with varying success in the attempts of isolation and cultivation of agents in the CC [198, 207, 369, 418]. Successful direct isolation of coxiellas in cell culture BGM (buffalo green monkey) after inoculation of clinical specimens from cattle, cats, dogs, and tick announced Willems et al. [553] and Ho et al. [244]. To increase the sensitivity of the cell cultures to infection with C. burnctii are sought various options— contamination by centrifugation [207, 369, 418, 486], improve culture broth, and others. These data indicate that in its natural state (without preliminary or additional impacts), the cell cultures are not as susceptible to infection with native suspensions of clinical and pathological materials in Q fever in humans and animals. In our attempts, we turned our attention to two CC—primary CEF (chicken embryo fibroblasts) and stable line BHK-21 (baby hamster kidney). We received very good results in the use of a new way of infecting the CC, namely inoculation with concentrated and purified suspensions (10^7–10^9) of sheep (abortion, mastitis) and bovine (endometritis, metritis, mastitis) origin. Under these conditions, we observed marked infection with C. burnetii with large accumulations of the agent and a cytopathic effect. Detailing of the results shows that it is achieved direct isolation of the agent in CEF from aborted sheep placenta, cotyledons, and fetuses inoculated in the form of concentrated and purified inocula with higher infectious titer as the appropriate starting materials are cotyledons, followed by the placental tissues and fetal internal parenchymal organs. In other successful experiments were used concentrated and purified suspensions C. burnetii multiplied in a CE after isolating them from similar ovine materials, as well as from milk exudates and uterine secretions of sheep and cows. We find that the described method is convenient and takes precedence over the above treatments and effects on cell cultures in order to cultivate C. burnetii. In our opinion, this is a more rational way of infecting cell cultures approximating to the greatest extent to the natural way of infection [597].

6

Virulence and Pathogenicity

Questions concerning pathogenic and virulence properties of C. burnetii are a constant subject of studies in the current history of the problem Q fever. Noshimson, 2004 [378] analyzes key facts about the pathogen and Q fever and concluded that C. burnetii is extremely virulent infectious agent. It was found that one gram of aborted sheep placenta may contain over 1 billion cells of the agent [509]. The pathogen easily becomes air borne in the form of very small droplets, aerosols, or dust to the environment and susceptible animals and humans [129, 231, 509]. In the experimental conditions, inhalation of one cell of C. burnetii may cause infection and clinical disease in humans [129, 355, 516]. According to the WHO, 50 kg of dried powder C. burnetii is capable of causing lesions at a rate equal to those of similar amounts of microorganisms that cause anthrax or tularemia [567]. The quality of highly infectious and extremely virulent pathogen is complemented by the ability of the agent to form spores-like form defining great resistance to inactivation and dominant aerosol pathway, especially during the birth period in animals [355]. For all these reasons, C. burnetii can be used as a means of biological terrorism [129, 163, 257, 378, 446].

Virulence properties of the agent have been studied on different biological models. There is evidence of high and low virulent isolates of C. burnetii from ticks Ixodes ricinus, caught in active foci of Q fever. Lowering the virulence of the agent in another type of tick (Dermacentor andersoni) by thermal effect (2°C) leads to the development in guinea pigs slight, inapparent infection [51]. In a previous study, Babudieri [85] concluded that circulates through ticks, C. burnetii maintain a high degree of virulence, while continuing living in the organisms of domestic ruminants leads to a spontaneous reduction in virulence. Pathogenicity of C. burnetii varies widely in respect of the species and nosological structure. Diseases in various forms and clinical manifestations are found in a range of domestic and wild mammals and birds, ticks, and humans. Natural and experimental conditions were obtained data on clinical and pathological consequences as a result of virulent properties and pathogenic action of C. burnetii of different origin. Despite a series of unexplained mechanisms, previous studies have shed light on the activity of the metabolism of major (LCVs) and small (SCVs) morphological forms of C. burnetii [346, 347] and have suggested hypothetical views on the ability of the pathogen to induce latent and persistent infection [87, 88, 126, 494, 510]. Some of the described too small cells (130–170 nm and smaller), resembling endospores [345], have the ability to pass through the filters. Suspensions, which pass through the filters of 100 nm, have pathogenic properties to chicken embryos and killing them [295]. The smallest cells of C. burnetii passing through the filters below 40 nm do not have infectivity and are non-pathogenic for CE [295]. Manifestations of pathogenicity and the subsequent lesions into white mice and guinea pigs are similar to those of human disease caused by C. burnetii, but vary depending on the mode of infection and possibly by immunological factors—the appearance of the first-phase and second-phase antibodies [124].

Khavkin and Tabibzadeh [276] and Hackstadt [223] investigated the role of LPS in virulence and pathogenicity action of C. burnetii. In the lungs of infected nasally white mice, they found that the damage to the endothelium of blood vessels is due more to the endo-toxicosis based on lipopolysaccharide of C. burnetii, rather than direct invasion of the agent [223, 288].

According to Ho et al. [244], isolated from these 59 strains coxiellas from dairy cows (raw milk, uterine secretions, and aborted fetuses) and ticks Ixodes sp., show three degrees of pathogenicity for guinea pigs: high (28/47.5%); moderate (23/39%); and low (8/13.6%). Protein and lipopolysaccharide (LPS) profiles of 12 of the above strains propagated in YS of CE and BGM cell culture were similar to those of the reference strain Nine Mile. LPS-established profiles of strains lead to the assumption that these agents are associated with acute Q fever [244]. Based on the DNA analysis, the strains were divided into two groups—inducing acute or chronic infections [239]. The above genomic groups I, II, and III are associated with C. burnetii, isolated from animals, ticks, or people with acute Q fever, caused by the so-called. Sharp strains. Isolates IV and V genomic groups associated with endocarditis in humans were referred to the so-called. Chronic shtamove. C. burnetii of VI genomic group isolated from wild rodents in Utah, USA, have not known pathogenicity. Virulence of different strains is explained more by the specific plasmid composition [455]. QpH1 plasmid found in the first, second, and third genomic groups was associated with "acute" strains of C. burnetii, and plasmid QpRS (fourth group)—with "chronic" strains [312, 454, 455].

The above-described hypothetical views of the virulence and pathogenicity of C. burnetii were not confirmed in broad surveys of a number of strains using RFLP analysis, LPS-analysis with specific monoclonal antibodies, and typing with plasmid specific primers [491, 513, 574]. Stein and Raoult [491] examined C. burnetii isolates of human origin and not found a specific gene responsible for pathogenicity. Masuzawa et al. [337] investigated the relationship between the pathogenic properties of 10 isolates (from aborted bovine serum, cow milk, human heart valve, human blood, and iksodovi ticks) and genes that determine their survival in macrophages (Cbmip—macrophage infectivity potentiator and qrsA—sensor-like protein). The authors present in >99% similarity between the Japanese, American, and European strains, but does not establish a link between their pathogenicity for guinea pigs and these genes [337]. Thiele and Willems [513] contest the validity of plasmid-based differentiation of C. burnetii of "acute" and "chronic" isolates. According to other beliefs, genetic variations of the strains C. burnetii obviously have a close relationship with the geographical source of isolates than with clinical manifestations of acute or chronic disease. The explanation for the development of an acute or chronic disease lies rather in the predisposing factors of the host, rather than the genomic differences of the strains [287, 574].

At this stage, it is not clear whether there is a correlation between the types MLVA (multilocus variable-number tandem repeat analysis) and virulence [604]. In the Netherlands, in the great epidemic of Q fever /2007–2010/ was found that only MLVA-type of the pathogen is responsible for abortions in goats in the country [605]. Obviously, a need for improved methods of molecular typing including developing genotyping with high resolution based on the sequencing of the entire genome and use thereof for testing of samples from humans, animals and from the external environment [604].

7

Antigens and Phase Variability

In earlier studies on the antigenic structure and antigenic properties of C. burnetii, it was found that the agent has two specific antigens—an antigen of the cell wall and somatic antigen [50, 501]. Under the influence of culture conditions, this microorganism occurs the phenomenon of "phase variation" [501]. Coxiella burnetii in nature (wild animals, ticks, domestic animals, and man) is in the so-called phase I, which is characterized by a high virulence and lower serological activity. After passages under laboratory conditions in chicken embryos or cell cultures, C. burnetii undergoes phase II, which is accompanied by a change in the antigenic composition, increasing the serological activity, a decrease in virulence, and the like. The pathogen in phase I has two antigens, of the cell wall and somatic, and phase II has only somatic antigen [50].

The phenomenon of phase variation in demand has different explanations. According to Stoker and Fiset [501] and Fiset [193], the reason for the passage of C. burnetii from second in the first phase are the specific antibodies against this agent produced by the organisms of animals and humans. Conversely, the absence of such antibodies in CE determines the reverse process—the passage of C. burnetii from first to the second phase. It is believed that the transition in phase II is a slow gradual process, varying in length for different strains of 7–40 passages. Returning from phase II to phase I is carried out in a much shorter time—2–4 passage in white mice or guinea pigs.

Other authors, however, expressed doubts about the role of antibodies to the phase state of C. burnetii [59, 108] and in general of the reversibility of the process phase variation [15, 271]. According to Baca and Paretsky [87], conversion from phase I to phase II is irreversible and is the result of a mutation caused by the deletion of part of chromosome structure. Similar findings strongly expressed in chromosomes found in attenuated isolates of the strain Nine Mile [538].

Hackstadt et al. [220] found in strains heterogeneity of structure and antigenic properties of lipopolysaccharide of C. burnetii in phase I. These findings in the LPS are due to mutations and are the bases of the phenomenon of phase variation [220–222]. Highly infectious agent in phase I is directly connected with lipopolysaccharide having smooth structure ("smooth LPS"). This phase is resistant to the action of complement. C. burnetii phase II is associated with lipopolysaccharide having a rough structure ("rough LPS"). Unlike phase I, phase II agent has damaged "crippled" LPS [74]—in addition, lacking some protein cell surface determinants [74]. There are differences with respect to the carbohydrate composition of the LPS in the two phases [75, 223, 463, 464]. In the first-phase koksieli, LPS contains sugars dehydrohydroxistreptosis, galactosamineuronyl-α- (1, 6) glucosamine, and L-virenosis which are not found in phase I of LPS.

Despite considerable progress, the overall decoding of the phase variation in C. burnetii from a genetic point of view remains as an unresolved problem.

Antigens of C. burnetii are determined using serological reactions, which will be considered separately. Using antigens prepared from single or pooled strains [409]. Perspective is the application of purified antigens. Sting et al. [497] receive the high cell-wall antigens by extraction with heat, ultrasound, SDS and phenol, and further purification by gel chromatography. After treatment of the antigens by proteases, lysozyme or meta-periyodat found complete retention of antigenic activity of the formulations, except the treated with meta-periyodat. The extracted heat and the gel-chromatography-purified antigens are used successfully for investigation into bovine sera in ELISA [497]. Hotta et al. 2004 [250] used monoclonal antibodies (MAbs) for analysis of the major antigens of C. burnetii. The authors establish the high specificity of the MAbs with the reactions of the proteins of the outer membrane of the agent, as well as with outer surface lipopolysaccharide. So, these components of the pathogen can be used as antigens for specific detection of antibodies against C. burnetii.

8

Serology

8.1 Serological Reactions

The diagnosis of Q fever in animals and humans in most cases is based on serological techniques in order to avoid difficulties in maintaining biological systems for the isolation and cultivation of the agent and the risks of intra-laboratory infections [118, 343, 355, 544].

For experimental purposes can be used different serological methods, but the most widely used are the complement fixation test (CFT), indirect immunofluorescence for the detection of antibody (IFA), and enzyme-linked immunosorbent assay—ELISA [54, 355, 378, 380, 409, 510, 533].

The complement fixation test is widely used in serological diagnostic practice for detection of antibodies, to differentiate and study of the newly isolated strains and for the titration antigens of C. burnetii. The main quality of the CFT is its high specificity. The sensitivity of CFT is moderately high but lower than that of IFA and ELISA [118, 395]—proving antibodies against antigens of C. burnetii in phases I and II. Used blood serum was obtained in the acute and convalescent phase of the disease. When the reaction is carried out with phase II antigen, a seroconversion is usually recorded 2–3 weeks after infection. Serological studies using CFT enable the differentiation of acute from chronic infections caused by C. burnetii [217, 395, 397]. For diagnostic purposes are widely used in phase II antigens. The use of antigens in phase I resulted in a positive shift reaction later in the re-convalescent period (30–40 days after inoculation), and maximum titers were reached at the 60th–70th day. Parallel investigations of blood sera of animals and humans with first-phase and second-phase antigens allow for following up the dynamics of epizootic process in foci of Q fever and to assess the freshness of the disease [31, 126, 284, 510]. In a recent message, Van Woerden et al., 2004 [533] presented results of a study of an epidemic of Q fever in humans in South Wales, UK, affecting 282 persons. Diagnosed serologically by CFT with antigen phase II. Tested 253 blood samples. Serological evidence of acute infection with C. burnetii (titers of 1:32 prevailing to 1:1024) was obtained in 95 cases (37.5%); doubtfully reacted are 42/16.6% (predominant value <1:8 in 35 patients, 2 to 1:8, 1:16, 1:32, 1:64 to 1); data for the previous infection were 8/3.1% (<1:8- 5 patients, 1:8- 3). In Germany [498, 499] have been conducted comparative serological testing of cattle and sheep by CFT and ELISA and have found good correlation and coincidence of positive results by both methods in 86.2%.

Essential for the proper functioning and effectiveness of the application of the CFT is the presence of qualitative antigens and investigation of well-processed and stored sera in order to avoid anticomplementary effects.

IFA method is also widely used, especially for serological diagnosis of Q fever in humans [129, 355, 391, 486, 518]. Some veterinary laboratories apply IFA for the detection of antibodies in ruminants [191, 228, 267, 532]. A positive reaction is established in the same terms as the CFT [23]. Sero-conversion establishes 10–15 days after infection [217, 395]. Often used are micro-process variants—micro-immunofluorescence test (MIFT)—which use small amounts of antigen. OIE indicates [409] that in order to exclude nonspe-cific fluorescence, the antigen does not need to be highly purified by repeated differential centrifugation. Suitable for the reaction are some commercial CF antigens [409]. Can also be used IFA-kits for the diagnosis of Q fever in humans as appropriate substitutions of the human anti-IgG fluorescent conjugate with bovine anti-IgG preparation [409]. Tissot Dupont et al. [518] describe MIFT using first-phase and second-phase antigens of C. burnetii— strain Nine Mile. Phase II antigen is derived from the pathogen, propagated in cell cultures—a first-phase antigen is prepared from the spleens of infected mice. Using the IFA may be determined antibody subclasses IgG, IgM and IgA. In the serological diagnosis of Q fever in animals for a positive response is taken sharpest immunofluorescence with serum dilution of 1:20 to 1:40. Values of approximately 1:160–≥1:320 can register upon acute infection or after a miscarriage caused by infection with C. burnetii [409]. Hatchette et al., 2001 [228] study in Canada abortions in goats and the associated epidemic of Q fever in humans (owners and farm workers and contact persons). In indirect immunofluorescence determine IgG antibodies against whole cells of Nine Mile strain of C. burnetii (antigens in phases I and II). For a minimum positive IgG titers accept ≥1:8. Acute infection with C. burnetii is characterized by a second phase of IgG antibodies titers of 1:64 or 4-fold increase in titer values for double study. In wide limits, titers in 82 (55.8%) out of 147 tested goats range from 1:8 to >1:4096 [228]. A parallel study of 179 people established the prevalence at 80/44.7% (second-phase IgG antibodies) as 66 of them (36.9%) had titers of 1:64 or seroconversion, indicating recent infection [228]. According to Marry and Pollak [327] in humans, the delimitation of negative titers depends on the immune background against C. burnetii in the study population, and therefore, there may be differences in different geographical areas. In acute Q fever, significant antibody titers considered ≥1:200 (second-phase IgG) and ≥1:50 (second-phase IgM). In chronic Q fever are detected antibodies against phase I antigen. First-phase IgG antibodies with values ≥1:800 are often an indication of endocarditis induced by C. burnetii [518]. In a recent publication, Madariaga et al., 2004 [308] describe in the USA endocarditis caused by C. burnetii in a patient infected with HIV. Etiological diagnosis is made serologically in IFA (IgG antibodies 1:16 384 phase I and phase II IgG antibodies, also 1:16 384) and by microscopic detection of C. burnetii in a heart valve [308]. Fournier et al. [195] suggested addition to Duke's criteria for the diagnosis of infectious endocarditis in humans, namely by including the isolation of C. burnetii from blood of patients or the presence of the first-phase IgG antibody (IFA) titer of ≥1:800.

Immunofluorescence tests for the detection of antibodies, especially MIFT, are easily attainable reaction achieving a saving of antigen and labor. Blood sera containing anticomplementary properties or hemolysis can be studied with this technique. For laboratories with fewer opportunities, there are restrictions to the introduction of the method, following the need for specialized equipment and regular supply with anti-species fluorescent sera.

The immunofluorescent hemocytic test /IHT/ test for the establishment of infection with C. burnetii of hemolymph of ticks is carried out using the method of Burgdorfer [116].

The enzyme-linked immunosorbent assay (ELISA) is described as being more sensitive than the IFA and CFT [263, 396, 409, 526, 542]. It is used in laboratories with the necessary equipment (a spectrophotometer, etc.). Find the first-phase and second-phase antibodies. Seroconversion is established generally between the 10th and the 15th day after onset of clinical signs [217, 397]. It takes relatively clean antigens. Some authors apply highly purified antigens [497]. To cover plates, wherein the reaction takes place can be used for antigens for CFT [409]. Jasper et al. [263] adapted competitive ELISA for the detection of antibodies against C. burnetii in the blood serum of cattle, sheep, goats, horses, and humans based on a monoclonal antibody with specificity for the lipopolysaccharide of the agent in combination with streptavidin peroxidase. Soliman et al. [481] also apply competitive EIA in goats, sheep and camels, and concluded good sensitivity and reliability of the method. In Germany, Dackau [151] develops capture-ELISA for detection of C. burnetii in milk samples as a possible alternative to the biological sample in guinea pigs. The test is based on monoclonal antibody and rabbit's immune serum against strain Nine Mile in the phase I. Experimentally infected with C. burnetii milk samples were processed by grinding with Pronase K and thermal effects with the aim of concentrating and inactivation of the antigen [151].

In its specificity, ELISA was inferior to CFT. Described are cross-reactions between certain strains of C. burnetii and Chlamydia [380]. Maurin and Raoult [343] indicate that the results obtained by ELISA are more difficult to interpret in comparison with the IFA. Therefore, ELISA is not widely used for the diagnosis of Q fever.

For serological diagnosis of Q fever in animals and humans have used a number of other tests—different variants of agglutination reaction, neutralization reaction, precipitation reaction, reaction opsonization, conglutination complement adsorbate reaction, and radioisotope precipitation test. Applied are also indirect haemolysis test [523], dot blotting [148], Western blotting [102] and others. For various reasons—sensitivity, insufficient specificity, labor intensity, complexity of implementation, the need for special equipment, or large quantities of ingredients—some of these reactions have been abandoned, while others are used rarely, mostly for experimental purposes. For example, conglutination complement adsorbate reaction is technically complicated and time-consuming; micro-agglutination consumes large amounts of antigen and is insufficiently sensitive and specific, unlike the newer version—high-density particle agglutination [379] and radioimmunoassay—although sensitivity and specificity can be used only in laboratories adapted for handling radioactive materials [168]. With respect to the immunoblotting techniques have been reported cross-reactions between certain strains Coxiella burnetii and Chlamydia [380]. Other authors have found Western blotting as too slow test with irreproducible results [343].

8.2 Serological Test's Examinations

8.2.1 Serological Data for the State of Agricultural Foci of Q Fever in Bulgaria

Table 8.1 presents the results of serological tests of domestic ruminants as a major agricultural reservoir of C. burnetii. It is seen that in the period 1977–1988 were tested 38,470 blood sera from cattle, sheep, and goats, and in

Table 8.1 Examination of sera of domestic animals for antibodies to C. burnetii for the period of 1950–2006

	Years		
Animal Species	1950–1976	1977–1988	1989–2006
Cattle	4749/939[a]	20086/2374	95737/5154
	19.77%	11.81%	5.38%
Sheep	17088/2856	16593/3123	99189/4764
	15.30%	18.22%	4.8%
Goats	1417/290	1791/193	54175/4123
	20.46%	10.77%	7.61%
TOTAL	23254/4085	38470/5690	249101/14041
	17.56%	14.79%	5.63%

[a] numerator—number of tested blood samples; denominator—number/% seropositive samples.

1989–2006 249,101 were collected from over 500 villages throughout the country. The results were compared retrospectively with data for distribution of Q fever in the initial period of the studies of the disease in Bulgaria. Three matched periods reflect different stages in the structure of the Bulgarian livestock [589, 597, 598].

Sero-epizootiology of the first period (1950–1976 years) reflects extensive livestock production by small private farms in its first half and created small- and medium-sized state-owned and cooperative livestock farms in its second half. The second period (1977–1988) is characterized by the introduction of industrial methods.

In the third period (1989–2006) occurs decentralization of agriculture and the creation of numerous fragmented small private farms and livestock farms.

Table 8.1 shows that Q fever is widespread among domestic ruminants, especially in the first period (1950–1976) and the second period (1977–1988), respectively, 19.77% and 11.81% in cattle; 15.30% and 18.22% in sheep; and 20.46% and 10.77% in goats. Studies in the third period (1989–2006) showed a reduction in seropositive cattle (5.38%), sheep (4.8%), and goats (7.61%). Within this 17-year period observed variations in size of seropositive reactions, respectively, extent of infection in years, animal species and in different geographical regions, districts, and farms [597, 598].

Table 8.2 reflects the results of serological tests for the period of 2002–2006 [598]. It shows that during those five years were tested 28,521 animals (15,866 cattle, 8727 sheep, and 3928 goats). Comparing the results show the highest rates of seropositivity in 2004, stand out data goats—21.65%, representing double increase compared with 2002 and three times compared with 2003.

Table 8.2 Serological examinations of domesticruminant animals for Q fever during the period of 2002–2006

Animal Species	Years					
	2002	2003	2004	2005	2006	обшо
Sheep	1819[a]	1811	1258	1911	1925	8727
	231/12.69%	151/8.33%	177/14.06%	291/15.22%	162/8.41%	1012/11.59%
Goats	677	1044	1016	832	359	3928
	80/11.81%	77/7.37%	220/21.65%	92/11.05%	69/19.22%	538/13.69%
Cattle	3006	3714	3188	3026	2932	15866
	247/8.21%	241/6.48%	310/9.72%	246/8.12%	310/10.57%	1354/8.53%
TOTAL	5562	6569	5462	5769	5216	28521
	558/10.03%	469/7.13%	707/13.07%	629/10.9%	541/10.47%	2904/10.18%

[a]numerator—number of tested blood samples and denominator—number/% seropositive samples.

8.3 Serological Status of the Infection Caused by Coxiella burnetii among Domestic Animals During the Epidemic of Q Fever in Humans in Etropole—2002

In April/May 2002 in the country, there has been a new upsurge in the incidence of Q fever in humans and animals in a defined geographical area—Sofia region, town Etropole and small villages in the immediate neighborhood [597, 598, 608].

However, epidemics have been affected in 123 people. They had a clinic sign with atypical pneumonia course. Infection with C. burnetii was laboratory-proven in more than 84% of patients. In this connection, it was determined the status of the infection among domestic ruminants in the directly affected region (Etropole), as well as in neighboring areas (Botevgrad, Trudovets, Osikovitsa, Litakovo, Pravets, etc.) in which it was not established Q fever in humans and in other parts of the country. The last two groups were controlling. The three groups: first—Etropole; second—adjacent areas; and third—other areas of the country, divided into two subgroups—samples tested to 20. 07. 2002 and other samples tested until September of that year, i.e., summer period (Table 8.3). This shows that CF antibodies to C. burnetii were found in 100% of the goats in 59.45% of the sheep and 11.62% of the cows or the average for the three animal species—25.42%. Titers of goats range from 1:8 to 1:512. Of these medium-high (1:32–1:64) and high (1:128–1:512) positive reactions were accounted for 86.6%. In sheep, the CF-titers are also in the range 1:8–1:512,—a medium-high (1:32–1:64) and high (1:128–1:512) were 72.09%. In cows, generally individual titers were lower: 1:8–1:16 (42.85%); 1:32–1:64 (39.28%); 1:256–1:512 (7.14%), or taken together, medium-high and high titers were 46.42%. During the same period, in the neighboring Etropole areas from 949 tested seropositive are 155 (16.33%). Here, positive reactions in goats are the highest—26.82%, and in sheep and the cows, respectively, 20.90% and the 10.68%. The country average of positive animals against C. burnetii in the period before 07.20.2002 was 12.84% (goats—18.32%, sheep—14.76%, and cattle—10.26%). Table 8.3 also shows that during the summer season (in this case arbitrarily assigned "after 20. 07. 2002"), the number of seropositive animals significantly reduced, especially in Etropole area where it has been active measures to curb the emerging epizootic and related epidemic (7.40%). A similar trend, but with less reduction, observed in the other two groups of regions (Table 8.3). The above data indicate the link between epidemic of Q fever in humans and farm outbreak of coxiellosis in the study area. For the unified etiology of the disease in animals and humans in Etropole also indicates the results presented in Table 8.4.

Table 8.3 Investigations into sera from domestic animals for antibodies to C. burnetii in 2002 (January–September)

Species	Areas									Total
	Etropole			Neighboring Areas			Other Areas of the Country			
Cattle	215/25[a] 11.62%	189/14[b] 7.40%	404/39[c] 9.56%	496/53 10.68%	152/4 2.63%	648/57 8.79%	1364/140 10.26%	1093/59 5.39%	2457/199 8.09%	3509/295 8.40%
Sheep	74/44 59.45%	–	74/44 59.45%	330/69 20.90%	154/32 20.77%	484/101 20.86%	948/140 14.76%	756/99 13.06%	1694/239 14.10%	2252/384 17.05%
Goats	6/6 100%	–	6/6 100%	123/33 26.82%	85/22 25.88%	208/55 26.44%	311/57 18.32%	210/15 7.14%	521/72 13.81%	735/133 18.09
TOTAL	295/75 25.42%	189/14 7.40%	484/89 18.38%	949/155 16.33%	391/58 14.81%	1340/213 15.89%	2623/337 12.84%	2059/173 8.40%	4672/510 10.91%	6496/812 12.50%

Numerator—number of studies; denominator—seropositive;
[a] group studied before 20.07.2002;
[b] group investigated after 20.07.2002;
[c] all.

Table 8.4 Serological tests for coxiellosis of domestic animals of Etropole belonging to sick people affected by Q fever (May 2002)

No. of Species	Sick Owner with Etiologically Proven Q Fever	CF-Titer in Blood Serum of the Animal
1. Goat	I. Ch.	>1:32<1:64
2. Goat	I. Ch.	1:64
3. Goat	S.C.	1:256
4. Goat	S.C.	1:512
5. Goat	S.C.	1:256
6. Goat	B.M.	1:128
7. Sheep	P.P.	1:64
8. Sheep	P.P.	1:32
9. Sheep	P.P.	1:128
10. Cow	P.P.	(−)
11. Cow	I. Ch.	1:8
12. Cow	I. Ch.	1:64

It can be seen that five patients with atypical pneumonia resulting from infection with C. burnetii are owners of seropositive for Q fever goats, sheep, and cows. Titers of CF–antibodies in goats range from >1:32 <1:64 to 1:512, in sheep from 1:32 to 1:128, while in cows–from 0 to 1:64. Positive serological results and the presence of active infection were confirmed by isolation of C. burnetii from the blood of goat No. 4 and sheep No. 9.

8.4 Serological Status of Infection with Coxiella burnetii among Domestic Animals During the Epidemic of Q Fever in Humans in the Region of Botevgrad—2004

In May 2004 in the region of Botevgrad, adjacent to Etropole, there is a new, larger outbreak of Q fever with affected of 220 people [597, 598, 606, 607]. On etiologically confirmed disease in humans corresponds to significant increase in seropositive animals C. burnetii in the same area: goats (63.33%), sheep (46.66%), and cattle (33%). Compared with 2002, the increase was three times that of cattle, almost two-and-a-half times in goats and more than doubled in sheep (Table 8.5). That finding is valid and comparing the results of Botevgrad in 2004 with the data for the presence of antibodies in animals in the neighboring regions of Botevgrad: goats (23.48%), sheep (18.29%), and cattle (16.43%)—an average of 18.86%, i.e., 2.6 times higher prevalence and other areas of the country through 2004, and: goats (13.48%), sheep (9.90%), and cattle (8.41%)—an average of 9.43 percent or 5, i.e., 2-fold greater (Table 8.5).

Table 8.5 Investigations into sera from domestic animals for antibodies to C. burnetii in 2004 (January–June)

Animal Species	Botevgrad	Neighboring Areas	Other Areas of the Country	Total
Cattle	100/33[a]	213/35	2875/242	3188/310
	33%	16.43%	8.41%	9.72%
Sheep	46.66%	164/30	989/98	1258/177
	46.66%	18.29%	9.90%	14.06%
Goats	150/95	132/31	734/94	1016/220
	63.33%	23.48%	13.48%	21.65%
Total	355/177	509/96	4598/434	5462/707
	49.85%	18.86%	9.43%	13.07%

[a] numerator—number of tested animals; denominator—seropositive animals.

In examining the exact geographical location of settlements in the Sofia region, where in 2002 and 2004 epidemics arise from Q fever shows that neighboring outbreaks of epidemic outbreak in 2004 (Botevgrad) are areas of Pravets and Etropole.

They surveyed a total of 213 cows (Table 8.5). Of these, 28 were from Pravets—an area free from clinical disease Q fever in humans—and the seropositive were 5/17.85%. The remaining 185 animals were from Etropole, and the seropositive cows—30/16.21%. Rates of antibodies against C. burnetii had similar levels in the bovine population in the group "Neighboring areas," and the average was 16.43% (Table 8.5).

The results of our studies indicate that Etropole—focus of epizootic and epidemic outbreak in 2002, two years later—maintains a level of antibodies against C. burnetii in cows, which is two times higher than that of the "other" regions of the country (8.41%), but is twice lower than the presence of antibodies in cattle in active focus of epizootic and epidemic of Q fever in Botevgrad.

Comparing the percentage of seropositive cattle in Etropole in 2004 (16.43%) with this in 2002 during the epidemic (11.62%) shows that in 2004 in Etropole there is wider circulation of infection in these animals.

The above (Table 8.3) high degree of infectivity of the sheep in Etropole in 2002—59.45% and goats—100% gives reason to believe that these two species animals are the main sources of infection with C. burnetii for the people in this epidemic [597].

Table 8.5 also shows that the regions located immediately adjacent to Botevgrad (Pravets, Etropole) in sheep detect antibodies against C. burnetii in 18.29%. And here, there is a trend similar to that observed in cattle: the level of seropositive reactions, which is two times higher than that in "other" areas of the country, but two-and-a-half times lower than that in active focus of Q fever in Botevgrad. Similar is the situation in goats (Table 8.5).

The only investigated cow whose owner is sick with pneumonia based on etiologically proven Q fever was seropositive for C. burnetii by CF—titer of 1:64.

Besides epidemic outbreaks and major epidemics among the population, Q fever occurs as sporadic, family or group diseases found in targeted serological tests. In some cases, evidence of the disease in humans predates the establishment of Q fever in animals. Such an example is given below.

8.5 Q Fever in Elenovo Village, Sliven Region

In this village in 2002, the outbreaks of Q fever in humans were found. Prior study on C. burnetii infection in animals was not carried out, but subsequently received information indicating the circulation of the agent and active coxiellosis among ruminants. It is also an infection of ticks R. bursa, downloaded from sheep to 30.55%. Table 8.6 shows the serological tests of nine animals owned by R.M.R. on November 20, 2012, whose family has two members affected by Q fever. It is seen that six sheep tested positive for C. burnetii at high or moderately expressed titers of CF antibodies. Of the three tested cattle, a single animal reacted with a low titer (1:16), and the others were negative.

In July 2004, from the same village, 15 blood samples from animals again in connection with cases of Q fever in humans (Table 8.7) were received. The table indicates total prevalence of test animals by 66.66% as goats in this group are the most resistant to C. burnetii infection with three negative and two weakly positive results. In cows and sheep is a significant prevalence (80%) and various titers, mainly in high and medium terms [597].

Table 8.6 C. burnetii antibodies in the blood sera of animals owned by R.M.R., whose family has been affected by Q fever (Elenovo village, Sliven region, November 2002)

No. of Sheep	CF-Titer	No. of Cattle	CF-Titer
1	1:512	1	Negative
2	1:256	2	Negative
3	1:64	3	1:16
4	1:128	–	–
5	1:64	–	–
6	1:32	–	–

Table 8.7 Serological testing for coxiellosis to ruminants from Elenovo, Sliven region, owned by sick people with Q fever (July 2004)

No. Cow	CF-Titer	No. Sheep	CF-Titer	No. Goats	CF-Titer
1	(−)	1	1:128	1	(−)
2	1:8	2	1:128	2	(−)
3	1:32	3	1:16	3	1:8
4	1:64	4	(−)	4	(−)
5	1:128	5	1:32	5	1:16

8.6 Serological Examinations in Sheep with Abortions, Stillbirth, and Deliveries of Non-viable Lambs

Our attention was focused on 56 farms from 20 areas where there were clinical manifestations of the pathology of pregnancy in sheep, often with mass characters: abortions, premature births, stillbirths, lambs born unviable that died within 48 hours. In many cases, serological tests were accompanied by microscopy, electron microscopy, and IF studies, as well as attempts at isolation of the pathogen from clinical and pathological materials of sheep and fetuses.

Upon serological examinations of 1008 aborted sheep is established that antibodies against C. burnetii were detected in 440 (43.65%) of them. Titer values ≥1:32 (1:32–1:1024) were found in 298 (67.72%) of the animals that reacted positively. Seropositive animals have found in 42 farms from 15 areas, of which 34 for the first time. In five areas have not received serological evidence of infection in aborted sheep. Serological tests were negative in 14 holdings [595].

Positively reacted aborted sheep were >50% in 26 farms. This positive serological response was between 10% and 40% in 16 farms.

Upon control serological investigation of 2,500 healthy animals—non-pregnant ewes in 5 holdings positive reactions were not found.

Figure 8.1 presents data from the statistical processing of the results [12]. Shown by percentage distribution of CF—titers between 10 and 30 days after the abortion or birth of dead and non-viable offspring shows that moderate values (1:32–1:64) are 53, 68%, and high-level antibody (1:128–1:1024)—46.29%. Geometric mean titer (GMT) is 91. The level of titers bears testimony to recent active infection with C. burnetii. The second part of Figure 8.1 contains data for the same indicators but in a later period—six months after the abortion or birth of dead and non-viable offsprings. In a significant percentage (59.28%), there is clearance of the titers. Antibodies against C. burnetii were detected at 40.72% of sheep, but 31.42% of them have low titers (<1:32) 1:4,

1:8, 1:16. Average levels of titers (1:32–1:64) is found in 9.27% of the animals. Geometric mean titer of the whole group of 140 sheep, including seronegative is <2 (Figure 8.1).

The control formed two other groups of animals: The first consisting of sheep had normal births in farms where there were abortions and other clinical conditions of pathology of pregnancy caused by C. burnetii,

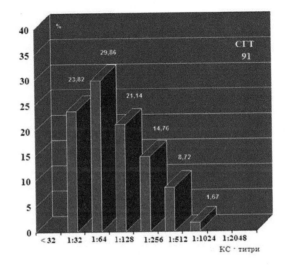

298 sheep

Abortions, stillbirths, non- viable lambs before 10–30 days

100 % seropositive; GMT 91

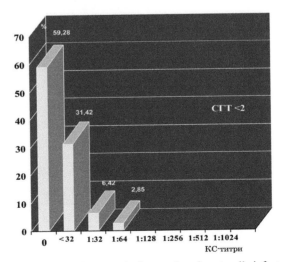

140 sheep

Abortions, stillbirths, non-viable lambs before 6 months

40,72 % seropositive; GMT < 2

Figure 8.1 Distribution of serological titers and geometrical mean titers in naturally infected with C. burnetii sheep with miscarriages, stillbirths, or births of non-viable lambs.

and second with sheep after normal births in farms which do not have abortions and others based on the Q fever (Figure 8.2). In the first control group of 90 sheep dominant negative (63.34%) and low titers (10.00%), while with values –1:32–1:128 react 26.66% of the sheep. The geometric mean titer for the group was 3.5. In the second control group, also consisting of 90 animals,

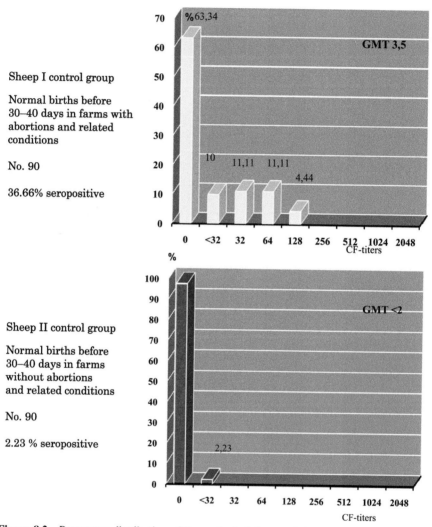

Figure 8.2 Percentage distribution of theserological titers and the geometric mean titer in sheep after a normal birth in farms without problems in terms of Q fever in comparison with other farms with established with Q fever.

the percentage of positive seroreagents is quite minimal, and GMT is <2 (Figure 8.2). The differences between the geometric mean titers in sheep with recent infection and the other three groups—aborted, stillbirths, and non-viable offspring—before 6 months and normal births from farms without a problem on the one hand and on farms having the above problems on the other are more than 3 (>3), indicating that they are reliable.

Our studies conducted on a large number of serum samples from numerous flocks of sheep showed that abortions associated with Q fever, the majority of animals have CF antibodies with different titers in the first 10–15 days after the abortion, and in some cases titers can do appear earlier (3–7 days). Serological diagnosis is facilitated when investigating a large number of aborted sheep from the same herd or from different herds in the same large farm, located territorially in the same place. In these cases, the detection of the CF antibodies against C. burnetii with different titers and levels gives rise to the assumption that abortion has etiology associated with Coxiella burnetii.

In a study of a single sample of blood from aborted sheep or a small number of samples from a group aborted animals considered 1:32 for a minimum positive titer suggestive of possible infection caused by a C. burnetii.

Except for abortions in Q fever above refers to premature births and cases of stillbirth and non-viable lambs at birth.

As an illustration of the possibilities of a single serological test to establish the involvement of C. burnetii in abortions and other similar conditions in sheep is presented in Figure 8.3. The chart shows the comparison of serological data in two groups of outbreaks. In the first group of 8 outbreaks, 13.12% of the sheep have a titer of 1:32 and 67.49% 1:64–1:1024. Low titers 1:8–1:16 are 19.37%. The geometric mean titer was significant—74. Opposite is the picture in the second group of 10 outbreaks where seronegative animals were 92.72% and in remaining 7.28% are detected low titers 1:4–1:8 (5.45%) and 1:16 (0.9%). Only at 1.36 percent CF-titers were 1:32 and GMT is <2. These data lead to the conclusion that in the first group, outbreaks with high probability C. burnetii is the causative agent of abortion and other related medical conditions.

The method of double testing of blood samples taken at intervals of 2–3 weeks was applied in cases when a stock or group of sheep in the first week after the abortion (premature births or stillbirth) is discovered massive negative or low titers of CF antibodies (1:4–1:16). To reliably growing titers (seroconversion) demonstrating the etiological role of C. burnetii in these clinical cases reported four times and greater rise in titer in the second sample compared with the titer of the first. For doubtful rise, we have assumed twofold

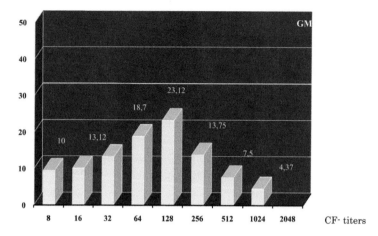

8 holdings; 160 sheep with abortions and related conditions; seropositive – 100%.

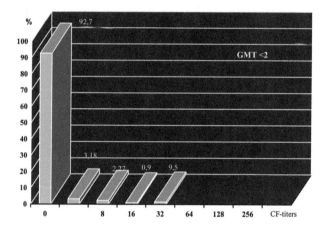

10 holdings; 220 sheep with abortions and related conditions; seropositive – 7.28%.

Figure 8.3 Comparative serological evidence for the etiologic role of C. burnetii at the outbreaks of miscarriage, premature birth, stillbirth or non-viable newborn lambs.

increase in titer. We are considering also the lack of difference in levels between the titers in the first and second samples and decrease in titer in the second sample compared with that of the first (inverse dynamics).

An example of twofold serology in abortions and stillbirths is presented in Figure 8.4. The graph shows that the object of the study were 42 sheep from 6 farms, where the first serum sample was taken between the 1st and 7th day

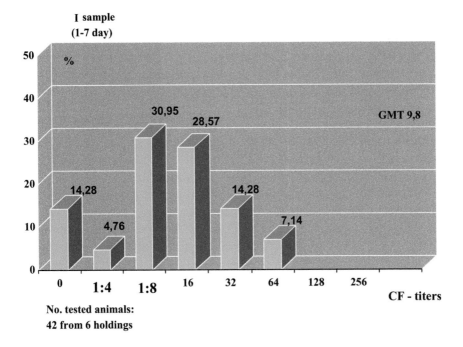

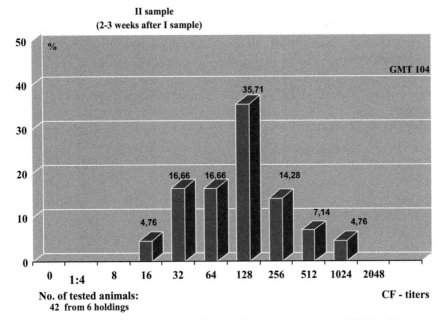

Figure 8.4 Double serological testing of sheep after miscarriages and stillbirths. Percentage distribution of the titers and geometric mean titers.

after the abortion (stillbirth). Antibodies against C. burnetii in this period are in the range of 1:4–1:64, and some of the animals were serologically negative. The geometric mean titer was 9.8. When examining the second serum sample titers range from 1:16 to 1:1024 as high values dominate (1:128–1:1024)—62%, while the GMT is 104. The difference between the GMT in the first and second testings was reliable (>3).

In cases where the first blood sample is taken at a later time—10–20 days after the abortion—the accumulation of antibodies is more significant and individual titers, respectively. GMT is higher. The same applies for the increase in titer and the GMT in the second serum sample.

Data from twofold examinations of sheep having pathology of pregnancy are too illustrative and demonstrative using the coordinate system (Figure 8.5). X-axis and y-axis register antibody titers in the first and second samples, respectively. Continuous median line outlines the projections of the titers, and parallel dashed line is the boundary of the fourfold increase in titers. The results of the double serological examination of each animal are represented as a point on the X-axis which is equal to the titer of the first test sample and the ordinate is equal to the titer obtained in the assay of the second sample.

It is obvious that in the graph, most of the results are over the median line and over the situated parallel line tracing the truthful increase in the titers. It is seen also that in five sheep, antibody titers in the second sample are decreased, and in the other three, the titers are obtained when the first and second studies are identical. These cases relate to the late taking of the first serum sample (30–40 days after an abortion or stillbirth) due to which the antibody level in it is significantly higher than in the second blood sample (Figure 8.5).

Completely opposite is the situation in other epizooties in sheep presenting with similar clinical symptoms (Figure 8.6). Double serological testing for coxiellosis showed no seroconversion and mostly equal low antibody titers in both studies or either loss of the serological response in the second sample. These results indicate that it is miscarriages and stillbirths of another etiological basis.

The data presented to serology on twofold blood samples surely prove the involvement of Coxiella burnetii in part of the abortions and stillbirths in sheep. The establishment of diagnostic seroconversion rate in the other related conditions—premature deliveries and non-viable offspring—can also be used to determine the etiological role of C. burnetii [595].

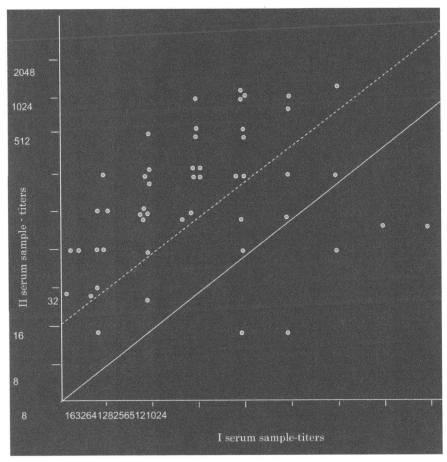

Figure 8.5 Graphic distribution of the results upon double serological examinations of 50 sheep after abortion and stillbirths with the etiological participation of Coxiella burnetii.

The duration of storage of complement-fixing antibodies to the antigen of strain Henzerling, situated in the phases I and II in naturally infected sheep of active foci of Q fever, is shown in Table 8.8. It is seen that the observations comprise remote time after the onset of infection—1 to 5 years. Antibodies to the antigen phase I stay 1½—2 years in low titers. Antibodies to phase II antigen are found in low and mean titers up to 1½ years, and in later periods, 3, 4, and 5 years, are kept at low levels in progressively decreasing number of animals (Table 8.8). These data should be considered when trying to retrospectively demonstrate the disease [597].

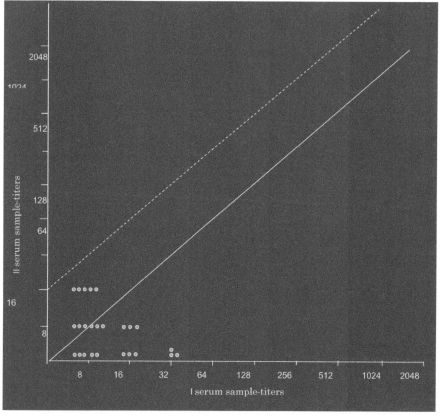

Figure 8.6 Graphic distribution of the results of the double serological examination of 25 sheep from outbreaks of abortion and stillbirths without involving Coxiella burnetii.

The dynamics of the titers of CF antibodies to C. burnetii was tracked in two groups of sheep after their *experimental infection* [610]. *The first group* included three pregnant ewes as follows: First at the beginning of the 4th month, second early in the third month, and third in the 45th day of pregnancy. The animals seronegative for Q fever, chlamydiosis, brucellosis, and toxoplasmosis were infected intravenously with Bulgarian strain "Chilnov" isolated from aborted sheep. Titration of antibodies in CFT was held with the antigen of C. burnetii phase II. The results are presented graphically in Figure 8.7. This shows that antibodies with titers of 1:8–1:32 appear on 8–10 days after infection. On the 30th day, the titers were grown to 1:64–1:128. The maximum titers (1:256–1:512) were established on the 40th day in two animals and on the 50th day in the third. These titers remained unchanged until the 60th day and then started to decline. The dynamics of the

8.6 Serological Examinations in Sheep with Abortions, Stillbirth, and Deliveries 97

Table 8.8 Duration of storage of CF antibodies in blood sera of naturally infected sheep to the antigens C. burnetii in phases I and II

		Complement Fixation Test											
		Antigen C. burnetii Phase II						Antigen C. burnetii Phase I					
		Positive		Titer				Positive		Titer			
Term of Research	Sample Tested /No./	No.	%	1:10	1:20	1:40	1:80	No.	%	1:10	1:20	1:40	1:80
12 months	7	7	100	1	2	2	2	3	42.8	2	1	–	–
18 months	7	6	85.7	2	2	1	1	2	28.5	2	–	–	–
2 years	7	5	71.4	2	2	1	–	1	14.2	1	–	–	–
3 years	7	3	42.8	2	1	–	–	–	–	–	–	–	–
4 years	7	2	28.5	2	–	–	–	–	–	–	–	–	–
5 years	7	1	14.2	1	–	–	–	–	–	–	–	–	–

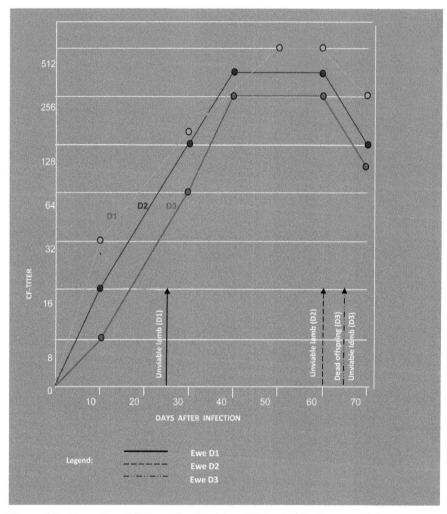

Figure 8.7   Dynamics of KC—antibodies (phase II) in the blood serum of three pregnant ewes following experimental infection with C. burnetii.

second-phase antibodies in the assay continued until day 70 after infection. The graph shows the outcome of infection with C. burnetii—the birth of puny non-viable lambs on 25th day after infection (ewe D1) or 60th day (ewe D2). In the third ewe (D3), infection led to the birth of twins—a dead lamb and a lamb unviable (Figure 8.7).

The second experimental infection of sheep also serologically negative for the aforementioned infections was achieved by i.v. inoculating the strain of C. burnetii in phase I, isolated from sheep. Animals were not pregnant. In this experiment, we tracked the dynamics of both the second-phase and the first-phase CF antibodies against C. burnetii, for a period of 18 months by repeated serological tests at intervals of 1 week to 3 months (Figure 8.8). In the titration of antibodies against the antigen II, phase pattern was observed as follows: appearance of antibodies during the first 7 days, with values

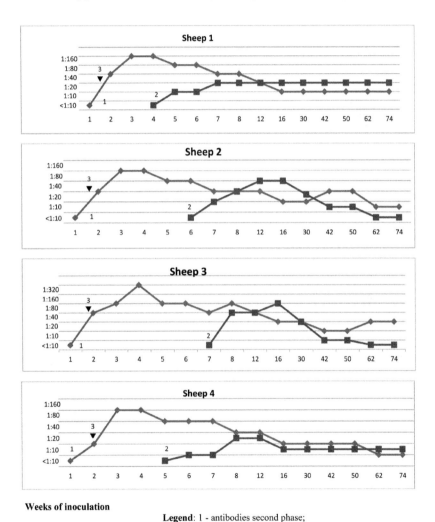

Weeks of inoculation

Legend: 1 - antibodies second phase;
2 - the first phase antibodies
3 ▼ - fever

Figure 8.8 Dynamics of the first-phase (2) and the second-phase (1) CF-antibodies of four sheep experimentally infected with C. burnetii.

1:10–1:20, a progressive increase in the second and third weeks (1:40–1:160), maximum levels (up to 1:320) in the fourth week, and keeping them up to six weeks (with sheep No. 3 to seventh) after infection. In later periods—from seventh (sheep No. 1, No. 2, No. 4) to 74th week (deadline experiment)—in these three animals are found the following dynamics of antibodies:

> sheep No. 1–1:40 (7–42 weeks), 1:20 (62–74 weeks)
> sheep No. 2–1:20 (7–12 weeks), 1:40 (16–30 weeks), 1:20 (42–50 weeks), 1:10 (62–74 weeks)
> sheep No. 4–1:20 (7–42 weeks), 1:10 (62–74 weeks).

Obviously, the detailed titers (and deadlines) testify to the prolonged detention of the second-phase antibodies in similar inverse dynamics including the level 1:40, 1:20, 1:10.

In this study, sheep No. 3 differs from the other three in terms of storage of a high titer: 1:160 in the seventh week, 1:80 in the eighth week, 1:40 in 12–16 minutes, and again in 50–74 weeks, which is the deadline monitoring.

Regarding the first-phase antibodies, we found significant differences. The appearance of antibodies in the serum was significantly later (between the fifth and seventh weeks after the infection). The level of emerging antibodies is too low: 1:10. Low titer 1:10 was registered between 7 and 8 weeks. Maximum titers ranging between 1:40 and 1:80 as they were established only in two sheep (No. 2 and No. 3) during the third–fourth month. These values are kept low and followed by a rapid decline to the levels 1:20, 1:10 and <1:10. Two of the experimental animals (No. 1 and No. 4) had low titers throughout the period of observation (<1:10, 1:10, 1:20) [610].

8.7 Serological Examinations of Rams

Subject to serological testing were 1662 rams originating from 30 farms and two quarantine facilities for imported animals. In six of the farms were established abortion and other clinical manifestations of the pathology of pregnancy related to C. burnetii and other twelve—Q fever without such clinical manifestations. It was found that in 56 rams (3.36%) were detected antibodies against C. burnetii. Serologically positive animals belonged to 15 farms, including those with abortions, stillbirths and deliveries of non-viable lambs. Positively reacted rams were discovered more in farms with cases of pneumonia etiologically associated with C. burnetii and also in latently infected herds where revealing dominant mildly flow and asymptomatic cases. In 15 farms and in both quarantine facilities lacked positive seroreagents. The percentage distribution of CF-titers and GMT in seropositive

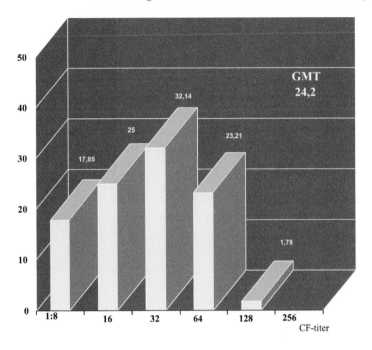

Number rams: 56

Figure 8.9 Percentage distribution of the titers and geometric mean titer in rams affected by Q fever.

for C. burnetii rams is presented graphically in Figure 8.9. Comparing the results for rams with serology of Q fever in sheep with established pathology of pregnancy showed obvious differences. Number of affected rams with antibodies against C. burnetii, the level of the individual titers, and their percentage distribution as well as the geometrical mean titer are significantly lower in rams. These data point to greater stability of males in ovine populations against infection with C. burnetii, respectively, for low prevalence of Q fever in rams [597].

8.8 Serological Examinations in Pneumonia in Sheep

These studies performed with samples from farms where they observed respiratory diseases among sheep and lambs [601].

Clinical, epidemiological, and laboratory evidence of infection and C. burnetii with similar symptoms got about 23 farms of 10 districts in northern, southern, and western Bulgaria. In some of the cases are known agricultural foci of animal coxiellosis where they found outbreaks of Q fever among people with mass character (Panagyurishte, Troyan, region of Kyustendil).

In the 18 farms, studied inflammation of the respiratory tract mainly affects older sheep. Serological characteristics were tested in 735 blood samples and 179 of them (24.35%) had antibodies against C. burnetii. The individual titers of the animals are in the range 1:8–1:512. The frequency of seropositive reactions to individual farms vary from 10 to 70%; in four, from 10 to 20%; in nine, from 30 to 40%; and in four, from 50 to 70%. Obviously, in these troubled farms with many cases of respiratory diseases in sheep was significant or high range of infection with C. burnetii detected.

The graphic image of the received serological results is presented in Figure 8.10. It is obvious the diversity of the serological titers, which is directly dependent on the stage of the infection in the individual animal, period in which the blood samples were taken, the possible antibiotic therapy

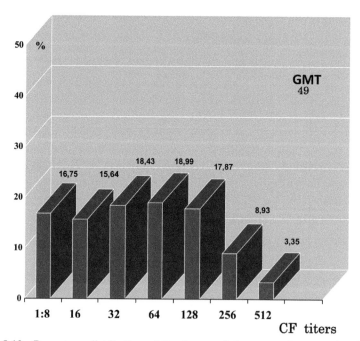

Seropositive animals (No.) 179

Figure 8.10 Percentage distribution of the titers and the geometric mean titer in ovine pneumonia with etiological participation of C. burnetii.

before making laboratory testing and so on. Generally, the ratio of the three groups titers: low (1:8–1:16), medium (1:32–1:64), and high (1:128–1:512), is approximately equal – by 1/3. The geometric mean titer was 49.

In some of the sick animals with pneumonia were tested twice serum samples—in the first week after clinical onset and 2–3 weeks later. Established seroconversion—a fourfold or greater increase in titers—shows the effectiveness of the method for serological diagnosis of pneumonia in sheep etiological associated with C. burnetii.

In five farms in five areas, pneumonia were observed in lambs aged 1 to 6 months, and in one of them, both in lambs of 12–14 months of age. In another holding was affected flock of lambs of the same age. From examined 94 animals, 49 (52.17%) responded positively against C. burnetii. The highest percentage of seropositive reactions found in older lambs – 71.4%. Lambs' – seropositive reagents on the same farm were 55.55% and in the other three, respectively, 60%, 37.5%, and 23.07%. Older lambs with pneumonia in the other farms were also seropositive for C. burnetti in a significant percentage (51.06%). Separate analysis of results for the lambs showed that 50% of the studied 40 animals with pneumonia have antibodies against C. burnetii. When tested 54 older lambs, the average rate of the reactants is approximately the same—53.7%.

Individual titers ranging from 1:8 to 1:256 and their percentage distribution are shown in Figure 8.11. This shows that in young and older lambs with inflammatory respiratory conditions prevailing average levels titers (1:32–1:64)—51%. High titer values (1:128–1:256) were 18.36% and low (1:8–1:16) 30.61%. The geometric mean titer was 42.

Comparison of the ratios of serological titers in lambs of different ages affected by pneumonia with those of older sheep with similar clinical characteristics shows similarity in low titers and prevalence of the average titers in young animals with about 18% and the prevalence of high titers in older sheep approximately by 17%.

Serological data for the etiological role of C. burnetii in pneumoniae in sheep and lambs received confirmation through isolation of strains of the pathogen from the lungs of infected animals that end in death. Surveys conducted for the chlamydiae, viruses, and pathogenic bacteria of specimens derived from the above-studied farms with pneumonia were negative.

Serological response in dynamics against C. burnetii, isolated in pneumonia in sheep studied in three experimentally infected animals—seronegative for Q fever in previously conducted research: two two-month-old lambs and an elderly non-pregnant sheep [601]. We used local isolate *Tr* from the lungs

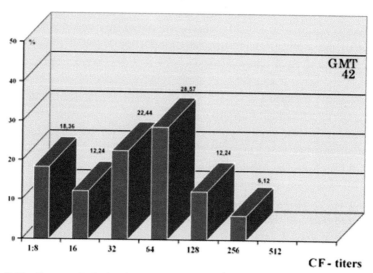

Figure 8.11   Pneumonia in lambs of different ages with the participation of C. burnetii. Percentage distribution of the titers and the geometric mean titer.

of lambs affected by naturally acquired infection with C. burnetii which is manifested by respiratory clinical signs. One of the lambs inoculated nasally, and the other lamb and sheep with tracheal infectious suspension of strain with a titer $10^6 ID_{50}$/ml, propagated in YS of CE. As a result of the infection, the lambs developed clinical picture of a general medical condition with distinct signs of inflammation of the respiratory tract. In sheep, the picture was similar but with more modest, subacute course. Were tested serologically blood samples from animals taken 7, 15, 30, 45, 60, 90, and 120 days after infection. The dynamics of antibodies against the C. burnetii is shown in Figure 8.12. It is seen that the most pronounced serological response gives sheep intratracheally infected, by the accumulation of specific antibodies with higher values of the individual titrations, the highest peak titer (1:512), and a slower decrease in the titer to the duration of observation (1:32 to 120 days). Lamb infected intratracheally also reacted with significant levels of antibodies, which are two times lower than in sheep, but two-fold higher titers than the lamb infected by nasal route (A1). In addition, a lamb A1 at the earliest showed seronegative reaction (Figure 8.12). Despite these differences, and clearly the greater efficiency of tracheal infection, in general, serologic response and dynamics of the antibodies in the three experimental animals are similar.

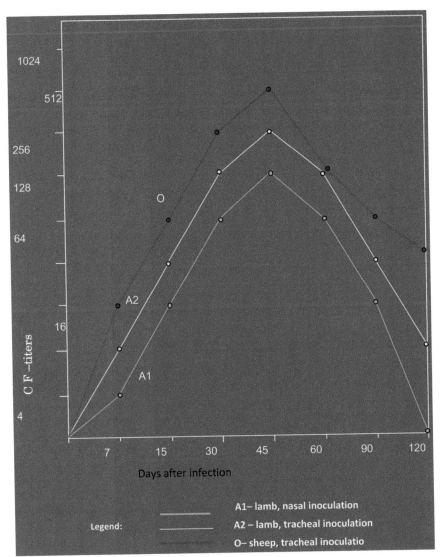

Figure 8.12 Serological response to lambs and sheep after nasal and tracheal infection with
C. burnetii isolate from the lungs of a lamb suffering from pneumonia.

8.9 Serological Investigations in Mastitis in Sheep

The first investigations on mastitis in sheep, caused by infection with
C. burnetii, were done in herds of northeastern Bulgaria, which had previously
been excluded contagious agalactia [602]. From four herds on the farm,

5(I–M), respectively, 18 (II–NV), 25 (III–SE) and 18 (IV–GD) blood samples were taken—a total of 66. Antibodies against C. burnetii (phase II antigen) were found in 30 (45.45%) animals with severe clinical form of mastitis (see Chapter 10).

The distribution of seropositive animals in herds was as follows: I herd—2 (40%), herd II—7 (38.88%), herd III—13 (52%), and herd IV—8 (44.44%).

Titers ranged from 1:8 to 1:512 as dominated by high levels (1:128–1:512), constituting 46.66% of positive reactions. Mean titer values (1:32–1:64) was 26.66%. The same proportion (26.66%) also had low titers (1:8–1:16).

These first serological data for Q fever in acute mastitis were convincingly supported by isolation of C. burnetii from strongly changed milky secretions of sheep and direct electron microscopic detection of the causative organism.

In another holding of the same area revealed mastitis caused by C. burnetii again by the method of complex etiological diagnosis—serology, isolation of the agent and direct EM. Serological surveys were conducted twice in 30 days. In the first study on 15 serum samples, 10 sheep with acute mastitis and 5 in contact with clinically healthy animals, 14 were seropositive (93.33%) with a level of individual titers and ratios of medium and low values similar to described above. High titers were not higher than 1:128 and accounted for 20% of positive reactions. Retesting after one month identified a significant serological response in all 15 sheep as a fourfold increase in titer was observed in four animals, including single seronegative sheep in the first titration. And in this outbreak of mastitis serological data for Q fever were confirmed by the culture and electron microscopic examinations of exudates from inflamed mammary glands.

In a third study of mass mastitis in sheep (Dragalevtsi, Sofia district), serological tests were performed with the second-phase and the first-phase antigens of C. burnetii. From 17 serum samples, 11 (64.70%) responded positively to an antigen phase I with low titers 1:8–4, 1:16–5, 1:32–2, and 10 (58.82%) and with a phase II antigen: 1:8–1, 1:32–4, 1:64–2, 1:128–2, and 1:256–1. These results suggest the involvement of C. burnetii as the etiologic agent of mastitis as evidenced by the isolation of the agent from milk secretion.

Pooled data from statistical processing of serological result in the three described mastitis outbreaks are exposed to Figure 8.13. From the distribution of the titers is seen that the titers $\geq$1:32 are 72.73% and $\geq$ 1:128 were 41.16% of positive reactions. The geometric mean titer was 49 [602].

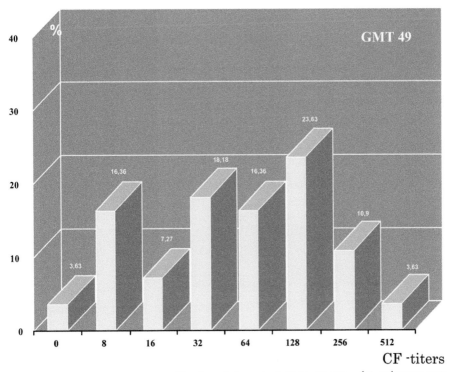

Figure 8.13 shows the percentage distribution

Figure 8.13 Mastitis in sheep. caused by C. burnetii. Percentage distribution of the titers and geometric mean titer.

8.10 Serological Examinations in Goats with Abortions, Stillbirth, and Deliveries of Non-viable Kids

We examined 1743 serum samples from goats with abortions, stillbirths, and unviable kids who died in the first 48 hours, coming from 18 settlements. The total number of seropositive for Q fever was 395 (22.66%). For individual not prosperous farms, extent of the positive reactions ranged from 15 to 66.66% [594].

The percentage distribution of titers in goats and GMT is illustrated in Figure 8.14. This shows that the highest titers (1:128–1:1024) were detected in 51.88% of the positive seroreagents. The detailed structure of these reactions is the following: 1:128–25.06%, 1:256–9.36%, 1:512–8.35%, and 1:1024–9.11%. The proportion of very high titer values is significant—17.46%. Medium expressed titers (1:32–1:64) constitute 30.89%

✦ With the participation of C. burnetii
Holdings (No.): 18;Positive (No.): 395

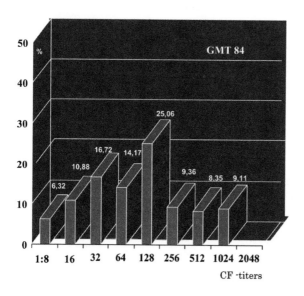

✦Without the participation of C. burnetii
*Holdings (No.): 16;*Examined goats (No.): 1488; *Positive (No./%):* 41/2.75%

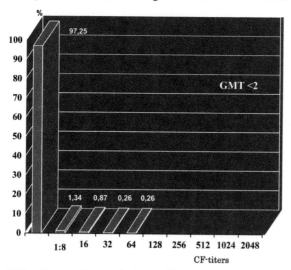

Figure 8.14 Comparative serological evidence of involvement of C. burnetii in the etiology of caprine abortions, stillbirth, and deliveries of unviable kids.

and the low (1:8–1:16) are 17.20%. The geometric mean titer was 84. The presented results attest to the recently passed active infection with C. burnetii, which is obviously etiological related to abortions and births of dead and non-viable kids. The involvement of the agent in the etiology of these conditions was convincingly supported by its discovery in placentas and fetuses via direct electron microscopy and isolation of strains.

The upper group outbreaks of abortion and other manifestations of pathology pregnancy were compared with a second group of diseases with similar clinical manifestations, including 1488 goats from 16 settlements. Antibodies against C. burnetii is found in 41 animals (2.75%) as 33 (2.21%) of them had titers of 1:8–1:16 and 8 (0.54%)–1:32–1:64. Seronegative were 1447 goats (97.25%). The geometric mean titer was <2 (Figure 8.14). These results show that in the second group of animals the abortions, stillbirths, and non-viable offspring are not related to infection by C. burnetii.

In farms with serologically proven Q fever were compared antibody titers in goats with abortions and other similar clinical conditions to those in contact with pregnant and gave birth normally goats in the preceding 30 days (Figure 8.15). Graphical data show that infection with C. burnetii is established in both groups contact animals prevail in coverage in group III (normal delivery)—72% and geometric mean titer of 19.7. In group II (pregnant), above indicators are, respectively, 64.84% and 14.9. The individual CFT titers in both groups are in the range 1:8–1:512.

When comparing the data for the three groups shows that GMT in animals with clinical picture manifested by abortions and births to non-viable or dead offspring (group I) was 5.6 times higher than GMT in pregnant goats (group II) and 42 times higher than GMT in goats with normal births (group III) [594].

In settlements with Q fever in goats accompanied by mass abortions and related conditions, we excluded infection with chlamydia, pathogenic bacteria, and contagious agalactia. As controls were examined, 850 clinically healthy goats imported from abroad, located in quarantine period. All animals were seronegative for C. burnetii.

8.11 Serological Investigations in Respiratory Diseases in Goats

The study included goats from 9 settlements in 6 regions of the country. Part of the farms were in known agricultural foci of Q fever where the infection is found in other types of domestic ruminants. Respiratory

Group I. Goats with abortions and related conditions (No.): 395

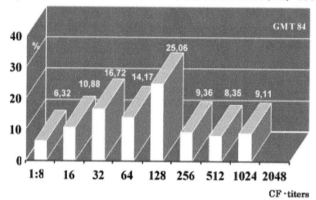

Group II. Contact pregnant goats (No.): 403

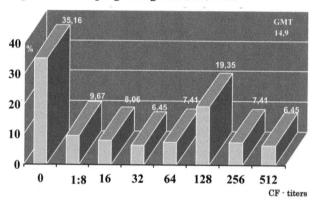

Group III. Contact goats with normal births before 15–30 days (No.): 300

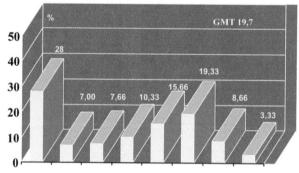

Figure 8.15   Q fever in goats. Percentage distribution of the titers and geometric mean titers in goats showing pathology of pregnancy and in contact animals.

conditions—pneumonia or acute bronchitis occurring more often as disease outbreaks and rarely as sporadic cases—affect all age group goats. In two flocks were observed simultaneously miscarriages associated with C. burnetii.

In the serological study of 249 adult goats and young goats, antibodies against C. burnetii are found in 103 (41.36%) [594]. Positive seroreagents to individual farms were from 22 to 100%, in four between 22 and 46% and in five between 47 and 100%. The average range of infection among females and males is approximately equal. The individual titers of the CF antibodies are in the range of 1:8–1:1024. Low titer values (1:8–1:16) are 16.50% and moderate (1:32–1:64) 32.03% of serological reactions. Accumulation of antibodies with high titers (1:128–1:1024) was found in 51.45% of seropositive animals. The geometric mean titer was 79. These data, shown graphically in Figure 8.16,

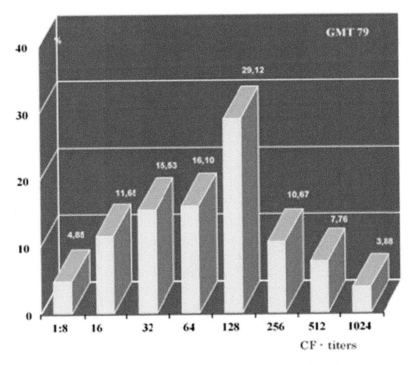

• *Settlements (No.): 9;* • *Examined goats (No.): 249;*

•*Seropositive for C. burnetii*

Figure 8.16 Respiratory diseases in goats caused by Coxiella burnettii. Percentage distribution of the titers and geometric mean titer.

talk about the high sensitivity of goats to inhalation infection with C. burnetii and expressed serological response in developed clinical picture of acute bronchitis or lung inflammation.

8.12 Serological Investigations in Bovine Miscarriages, Stillbirths, and Births to Non-viable Calves

These surveys cover 1395 blood samples from 102 settlements in 20 regions of the country. Serological evidence of infection with C. burnetii was obtained from 278 animals (19.92%) of farms in 45 villages from different regions and areas. In 57 settlements did not disclose specific antibodies in cows with abortions, stillbirths, or non-viable calves [591]. The percentage of serologically positive cows for individual farms vary from 0.32% to 100%. The results allowed us to distribute the incidence of abortion and related clinical conditions into three groups (Figures 8.17 and 8.18).

The first group includes abortions from the settlements where more than 50% of the tested cows were seropositive for C. burnetii. It is evident that the group includes miscarriages and related conditions found in 14 villages in seven districts.

The second group included abortions in 29 settlements of 14 regions where the positive seroreagents are 10–49%.

The third group are those cases of miscarriages, where serologically positive were less than 10% of the cows.

The serological examinations of the first group showed that 79 (61.24%) of the surveyed 129 cows react positively against C. burnetii with the CF-titers of 1:16 to 1:1024 (Figure 8.17). The geometric mean titer in positive responders is 84. The most frequently observed individual titer was 1:64 (34.17%), followed by 1:128 (29.11%). The share of high titer values (1:128–1:1024) is 45.55%, the average (1:32–1:64) 49.35% and the low (1:16) 5.06%.

Serological results obtained indicate the etiological role of C. burnetii in the cases of miscarriages, births of non-viable, or dead offspring which is confirmed by direct EM and isolation of the agent from placentas and fetuses. Moreover, apparently it concerns abortions and related conditions with a dominant involvement of C. burnetii.

Different is the situation in the second group. Figure 8.17 shows that in the assay of 646 cows at 69.51% were not detected specific antibodies agains C. burnetii. Seropositive animals (30.49%) are twice less compared with the first group. Titers were in the range 1:8–1:512. The distribution of high, medium and low titers is, respectively, 36.02%, 49.23%, and 14.71%. The geometric mean titer in this group of animals is significant—60. These results

I. Abortions and births of dead or non-viable calves of dominant etiological participation of C. burnetii

• *Settlements (No.): 14;* • *Tested cows (No.): 129;* • *Positive (No./ %):79/61,24%*

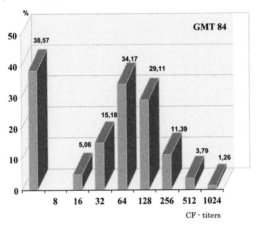

II. Abortions and births of dead and non-viable calves with partial etiological involvement of C. burnetii

• *Settlements (No.): 29;* • *Tested cows (No.); 646;* • *Positive (No./ %):197/30,4*

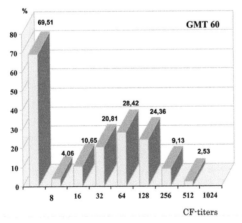

Figure 8.17 Percentage distribution of the titers and geometric mean titers in cases of miscarriages, stillbirth, and deliveries of non-viable calves.

are an indication of the partial participation of C. burnetii in the etiology of abortion and other forms of pathology of pregnancy in cows in the 29 farms surveyed.

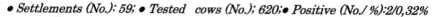

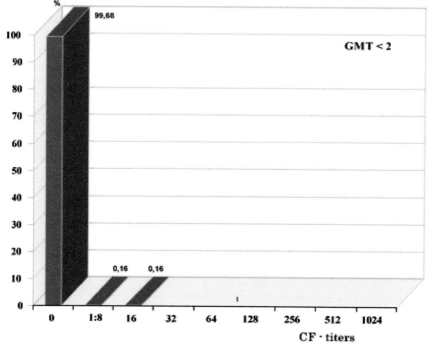

Figure 8.18   Abortions in cows and births of dead or non-viable calves without etiologic involvement of C. burnetii.

In the third group, including 620 cows from 59 holdings, 99.68% of the animals were serologically negative for Q fever (Figure 8.18). Only in the 2 cows (0.32%) were found CF antibodies with low titers—1:8–1:16. They came from two farms in the two areas from which examined 28 common serum samples. In 57 settlements with 592 cows tested positive, reactions lacked. These results exclude the participation of C. burnetii in the etiology of abortions and stillbirths in the surveyed farms [591].

8.13 Comparative Serological Examinations of Cows by CFT and MIFT

In a study on the spread of Q fever in the region of Haskovo, we tested serologically by complement fixation test (CFT) and the method of immunofluorescent titration the serum (MIFT) of 300 cows from three farms having problems with regard to abortions, retained placenta, and endometritis. Serologically positive for Q fever were 28% of the test animals [591].

A comparison of the results of the simultaneous application of both methods is presented in Table 8.9. The table shows that the CFT gives higher detection rate—80.95 percent against 71.42 percent for MIFT. The coincidence of the results of the two tests is 52.38%, but it appears that a certain percentage of seropositive animals are found alone in only one or the other method. Titers in CFT are from 1:10 to 1:80 (1:10–27/39.7%; 1:20–24/35.29%; 1:40–9/13.23%; 1:80–8/11.76%). In MIFT were discovered antibodies with titer of 1:10 in 30 animals (41.11%) 1:20–26/38.23%; 1:40–7/10.29%; and 1:80–5/8.35%.

The data obtained show that the MIFT has a certain diagnostic value and can be used for serological diagnosis of Q fever in cattle, the benefit of which is relatively light setting of the reaction and the visibility of results. Most secure, however, are the results obtained in the simultaneous use of CFT and MIFT as the correlation between the two tests is not constant and is probably based on the phase of infection [587, 591].

8.14 Serological Response in Cattle after Experimental Infection with C. burnetii

Dynamics in the titers of CF antibodies against C. burnetii tracked in two experiments with cows after their experimental infection [610].

First experiment. Two non-pregnant cows seronegative for Q fever in preliminary examination inoculated intravenously with infectious suspension (10^8) of C. burnetii in YS of CE infected with strain BP (Phase I) isolated from sheep at a dose of 10 ml. The strain was passaged in advance in white mice (8 consecutive passages) and cultivated again in CE by two consecutive passages. Serological investigations with phase II antigen started two days

Table 8.9 Comparative serological testing for Q fever by CFT and MIFT of cows with problems about the abortions, retained placenta, and endometritis

Seropositive Animals (Total No.)	Positive Simultaneously in CFT and MIFT No./%	Positive Only in CFT No./%	Positive Only in MIFT No./%	All Positive by CFT No./%	All Positive by MIFT No./%
84	44/52.38%	24/28.57%	16/19.04%	68/80.95%	60/71.42%

after inoculation and continued at intervals of 2, and later on from 5 to 10–20 days for 160 days. The results are presented graphically in Figure 8.19. Shown are the serological response curve and the time of appearance and retention of febrile period.

In cow No. 1, antibodies titer of 1:10 appeared on the 8th day and remained at that level until the 15th day. Follows a progressive increase in titer to 1:160 for 30 days, a reduction of one dilution (1:80) of the 40th day and rebounds to 1:160 of the 50th. After this period began the process of lowering titer: 1:40 (60 its day), 1/20 (70 its day), 1:10 (80 days), <1/10 (100 days). Later (140 its day) noted a slight increase in antibodies (1:10) and held at that level until the end of the experiment (160 its day). The graph shows also that the temperature rise began on day 4 after infection and continued until day 6, which precedes the appearance of antibodies. We note that at this cow had a relapse of fever, on the 120th day that preceded the above-described repeated, but a slight increase in antibody titer.

In cow No. 2, the febrile phase started on day 6 and lasted until the 8th day, and a week later (15-d day) appeared antibodies (1:10), was kept at this

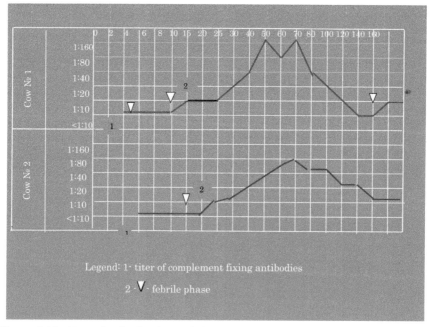

Figure 8.19 Dynamics in the titers of CF antibodies in the blood serum of two cows experimentally infected with C. burnetii.

level until the 20th day and then increased without fluctuations to 1:80 on day 40. Ten days later (50th its day), we noted a gradual decrease in titer to 1:10 on the 120th day and detention at this level until the deadline of observation.

Second experiment. The second experimental infection of cattle also achieved by intravenous inoculation of C. burnetii in phase I, isolated from sheep. For 74 weeks, we followed the dynamics of the first-phase and second-phase antibodies by repeated serological tests (Figure 8.20). Infection resulted in both cows with increase in body temperature on the 8th day (cow No. 1) and on day 12 (cow No. 2). In blood serum of two animals accumulated KC antibodies, which was manifested with pronounced dynamics. In No. 1 cow found occurrence of the second-phase antibodies (1:10> 1:10 <1:20) on the 13th–15th days, gradually increasing to 1:160 of the fifth week (in cow No. 1, that level was remained until the seventh week (after a temporary decline to >1:80 <1:160 6th week)) and inverse dynamics—gradually decreasing titer values between 7 and 16th weeks to a level <1:10–1:10. New growth of antibodies to > 1:20 < 1:40 was noted at 16–30 weeks, followed by retention of these titers and a decrease to 1:20 at the end of the experiment.

In case of cow No. 2, the maximum titer of 1:80 was noted at the sixth week. The same persisted for seven days and followed a gradual descent

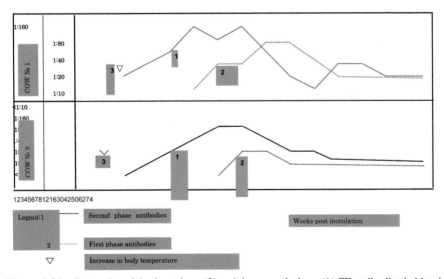

Figure 8.20 Dynamics of the first-phase (2) and the second-phase (1) CF antibodies in blood serum of two cows experimentally infected with C. burnetii.

to >1:20 <1:40 (12–16 weeks) and >1:10 <20 (30th). The final titer value persists from 30th week to 74th week (deadline of experience).

The dynamics of the first-phase antibodies in both animals is different. They appear later—fifth (cow No. 1)–6th week (cow No. 2) with a titer <1:10 to 1:10, soaring to 1:20 (6–7 week), and remain at that level about 7 days. At the first cow registered an increase to >1:40 <1:80 (8–12 weeks), gradually reducing to >1:10 <1:20 (30th week), and remained at that level until the 74th week.

At the second cow on the 12th week was the titer >1:10 <1:20 and the latter is maintained until the end of the experiment (74 weeks).

8.15 Persistence of Antibodies in Naturally Infected Cattle

The duration of storage of CF antibodies against the first-phase and second-phase antigens of Henzerling strain in cattle of active foci of Q fever is shown in Table 8.10.

The table indicates that the period of observation was 5 years as titration of antibodies is performed annually and once with the exception of the period between the first and second years, when it is in six months.

Antibodies to the first-phase antigen persist in some of the animals up to 2 years after infection in 50% (1 year), 33.3% (18 months), and 8.3% (two years). The titers are low—1:10.

Antibodies to phase II antigen persist in low and middle titers to four years in progressively decreasing number of animals, and in 2 cases—up to 5 years in low titer. These data are useful for retrospective diagnosis and sero-epidemiological analyses of Q fever in cattle [610, 597].

8.16 Serological Examinations in Pneumonia in Calves

Serological results of 163 calves with pneumonia originating from six settlements in four areas are presented graphically in Figure 8.21. It is seen that antibodies against C. burnetii were detected in 24.53% of animals. Seropositive reagents were found in the six villages studied, four of which had pre-recorded information about Q fever in cattle, sheep, and goats. In individual farms, extent of infection is different—from 18 to 50% [591].

The total percentage of medium–high and high titers (1:32–1:512) is very high–87.5%, and the ratio of the two groups' titer levels is approximately 1:1. Single animals react with lower titers. The geometric mean titer was 69. The results speak for recently spent active infection with Coxiella burnetii which should always be considered as a possible etiological factor for pneumonia in calves, especially in regions with established agricultural foci of Q fever.

Table 8.10 Duration of storage of CF antibodies in blood serum of naturally infected cattle to first-phase and second-phase antigens of Coxiella burnetii

Term of Research	No. Samples Tested	Complement Fixation Test											
		Antigen C. burnetii Phase II						Antigen C. burnetii Phase I					
		Positive		Titer				Positive		Titer			
		No.	%	1:10	1:20	1:40	1:80	No.	%	1:10	1:20	1:40	1:80
1 year	12	12	100	1	5	4	2	6	50	6	–	–	–
18 months	12	11	91.6	–	6	3	2	4	33.3	4	–	–	–
2 years	12	9	75	1	3	4	1	1	8.3	1	–	–	–
3 years	12	7	58.3	2	3	2	–	–	–	–	–	–	–
4 years	11	5	45.4	2	2	1	–	–	–	–	–	–	–
5 years	9	2	22.2	2	–	–	–	–	–	–	–	–	–

● Settlements (No.): 6; ● Tested calves; ● Positive (No./%): 40/24.53%

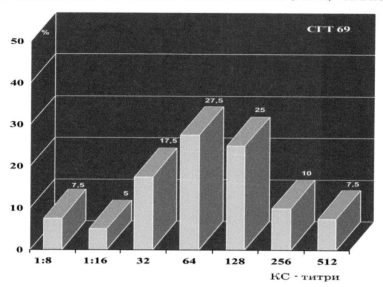

Figure 8.21   Pneumonia in calves etiologically associated with C. burnetii. Percentage distribution of titers and geometric mean titer of the specific antibodies.

In farms with serological evidence of inflammation of the lungs in calves caused by C. burnetii, the investigations for other etiologic agents were negative.

8.17 Serological Testing in Cows from Herds with Problems of Infertility and Decreased Milk Production

In this study, blood samples were taken from cows representing heterogeneous group of different clinical conditions and at different days of their discovery—sterility (common unsuccessful artificial insemination—the so-called repeats), inflammation of the reproductive system, the most common chronic puerperal endometritis, metritis, pyometra, retained placenta, mastitis [591]. In some cases, reducing milk production was a natural consequence of mentioned diseases that stand in the healthcare clinical status of herds and individual animals. In other cases dominated reducing the amount of milk until clinical signs were missing or were less pronounced. The last evidence for the presence of latent or oligo-symptomatic infection caused by C. burnetti. Serological results of 1,498 cows from 16 settlements in four areas

- *Settlements (No.):16*
- *Tested cows (No.):1498*
- *Positive (No./%): 335 /22.36*

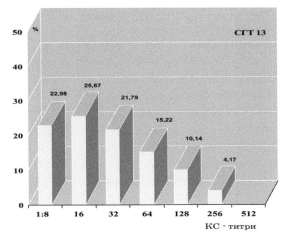

Figure 8.22 Distribution of titers and geometric mean titer in cows from herds having problems regarding infertility and decreased milk production.

are shown in Figure 8.22. It is seen that antibodies against C. burnetii were detected in 22.36% of tested animals, whether originating from settlements with previously unknown serostatus of the disease and from the known foci of Q fever, for example. Panagyurishte (Pz) and Piperevo (Kn) with fixed and massive outbreaks in animals and numerous cases of the disease in humans. In statistical processing of the data found the proportion of low titers on the one hand (48.65%) and the sum of medium and high, on the other (51.35%), with a slight majority (2.7%) in favor of the latter. However, in general, percentages of low and medium titer levels are significant, which determines the level of the GMT–13 [591].

8.18 Serological Examinations of Bulls

We investigated a relatively limited number of serum samples from bulls provided by some livestock breeding centers. In one of them (Yb) of 26 tested, one bull is serologically positive with CF-titer 1:64 (3.84%). In another center (Pd), 7 (13.46%) of 52 bulls had antibodies against C. burnetii: 1:16–1; 1:32–2; 1:64–3; 1:128–1. Clearly, the latter result indicates the presence of active infection with C. burnetii despite the absence of clinical signs expressed. These figures attest to the importance of serological tests for the detection of Q fever in bulls in the livestock breeding centers, which can be formed as foci of infection and source of infection for other animals [591].

8.19 Serological Examinations of Buffaloes

We examined a total of 1880 serum samples of buffaloes originating from ten farms in northern and southern Bulgaria. The survey included different groups of animals: buffalo cows, buffalo heifers, buffalo calves aged 2–6 months, and buffalo bulls 12–18 months of age [597].

Most of them were clinically healthy. Part of the heifers were aborted, and some of the dry pregnant buffaloes had pneumonia and conjunctivitis.

Antibodies against C. burnetii were found in 27 clinically healthy animals (1.43%) from three farms in three areas: 10 buffalo heifers, 4 female buffalo calves, and 2 buffalo steers with titers 1:8–7 (25.92%), 1:16–8 (29.62%), 1:32–8 (29.62%), 1:64–3 (11.11%), and 1:128 to 1 (3.7%).

Buffalos showing any clinical signs of disease were serologically negative for Q fever. In seven of the ten holdings, positive seroreagents were not found.

The data presented evidence of missing or poor circulation of the agent in flocks tested, respectively, of the significant resistance of buffalo populations to infection with C. burnetii [597].

8.20 Serological Testing of Pigs

The studies included 2514 sows from different regions of the country. The investigations were carried out in three periods: 1998–2004 (204 animals); 1984–1988 (1809); 1977–1983 (501).

Found only two serologically positive pigs—one in 1977–1983 (0.19%) and one in 1984–1988 (0.05%). Over the past 16 years, seropositive reagents have not been established. Obviously, the susceptibility of the population of pigs in the country to infection with C. burnetii is quite insignificant and at this stage it cannot be considered that the species is involved in the formation of agricultural foci of Q fever [589, 597].

8.21 Serological Testing of Horses

The results of these studies are shown in Table 8.11.

The table demonstrates that systematically conducted seroepidemiological studies of the infection in the horses in Bulgaria revealed a negligible percentage of positive seroreagents. Comparing data from the past few years with two preceding periods showed durability in the tendency of detection of single seropositive horses—a very low prevalence of the disease in this species [589, 597].

Table 8.11 Complement-fixing antibodies against C. burnetii in blood sera of horses

Years	1977–1983	1984–1988	1989–2004
Tested	389	445	450
Positive (No./%)	4/1.02%	4/0.89%	4/0.66%

8.22 Serological Testing of Dogs

We examined two categories of dogs, shepherds, and yard in rural foci of Q fever formed by ruminants and ownerless stray dogs in Sofia, captured and housed in metropolitan isolators [592].

The results obtained are presented in Table 8.12.

It is seen that the total percentage of serologically positive for Q fever dogs is considerable—43.93%. In active farm disease outbreaks, sheepdogs are in constant contact with both the infected sheep, goats, and cattle, especially when entering the pasture and with the main natural reservoir and vector ticks. Yard dogs in these areas are also in close contact with domestic ruminants reared in small private farms. Into this category dogs, the positive seroreagents are 49.45%.

Of particular interest are the data for antibodies against C. burnetii among stray dogs in Sofia, which establishes for the first time [592]. In this category of animals, seropositive reactions were 23.46% (Table 8.12). Individual titers ranging from 1:8 to 1:256 as the lowest value (1:8–1:16) constitutes 17.39%, the average (1:32–1:64)—69.56%, and higher (1:128–1:256)—13.04%. The geometric mean titer was 39.

8.23 Serological Survey of Cats

We tested serologically 32 domestic cats from two categories: home-grown and homeless wandering freely. The first group included both cats kept in almost closed environment of urban homes and cats from rural homes with

Table 8.12 Complement-fixing antibodies against C. burnetii in the blood serum of dogs

Category Dogs	Number Tested	Seropositive (No./%)
Shepherd dogs and yard dogs from active agricultural foci of Q fever	364	180/49.45%
Homeless stray dogs in the city of Sofia	98	23/23.46%
Total	462	203/43.93%

Table 8.13 Complement-fixing antibodies against C. burnetii in the blood serum of domestic cats

Category Cats	Number Tested	Seropositive (No./%)
Cats reared in homes	15	4/26.66%
Stray cats wandering	17	8/47.05%
Total	32	12/37.5%

yards and opportunities for movement within and neighboring zones. Captured stray cats were mostly from the city of Sofia and its vicinity [592].

Conducted serological examinations revealed a general prevalence of Q fever of 37.5% (Table 8.13) Higher amount of positive reactions is found in stray cats wandering (47.05%) Four serologically positive animals in the first category (26.66%) were grown in both rural homes (two cats) and in Sofia, also two cats.

8.23.1 Serology of Poultry

We directed our attention to several kinds of poultry—chickens, turkeys, and waterfowl, of which examined a total of 1997 blood sera (Table 8.14). The birds came from 16 farms in different regions of the country. Complement-fixing antibodies against C. burnetii with low titers (1:8–1:32) were found in chickens and geese, respectively, 4.24% and 4.09%. In 2005, for the first time in the country we found serologically positive ducks in a farm (5.6%)—Brezovo (Pd) [597].

8.24 Serological Examinations of Humans

8.24.1 Q Fever in Persons Professionally Engaged in Animal Husbandry and Veterinary Medicine

Our observations show that the prevalence of Q fever in persons working on farms that are problematic in terms of disease is common. The same goes for owners affected by Q fever goats, sheep, and cows kept in private yards,

Table 8.14   Serological testing of poultry for Q fever

Species	No. Tested	Positive (No./%)
Hens	1014	43/4.24%
Geese	366	15/4.09%
Ducks	445	25/5.6%
Turkeys	172	0

but sent daily grazing in mixed flocks. Examples of the relationship between diseases in animals caused by C. burnetii and their owners, presented in the same chapter on research in the settlements Etropole, Botevgrad, and Elenovo. The clinical forms of course are various—light, medium, and heavy, occurring mainly as a febrile illness with or without additional complications. Particularly sensitive are the people who fall for the first time in the contaminated environment—new workers, casual visitors, and others.

There are also latent asymptomatic forms that are found serologically.

Table 8.15 reflects the results of serological tests of shepherds and animals in the breeding herd of them. It is seen that people DI and P. G. react positively with titers of 1:64 and 1:128. In their herd No. 1 established infection with C. burnetii with higher prevalence (71.73%) and clinically mass abortions. The relationship between serologically positive reaction in sheep and the serving staff is obvious [597]. The situation is similar in the six working shepherds with flocks No. 2 and No. 3 of the same holding. Positive for C. burnetii titers ranging from 1:8 to 1:128, but values prevail 1:64–1:128. In ewes and adult sheep—a total of 70—there is an infection with C. burnetii of considerable scope—an average of 44.28% and clinical manifestations— miscarriages, premature and stillbirths, non-viable offspring, and retention of placentas.

Table 8.15 Serological data for Q fever in sheep and the shepherds of their herds (Studena village, Pernik District)

Herd No.; No. Sheep Tested; Clinical Manifestations	Positivesheep (No./%)	Percentage Distribution of the CF-Titers in Animals	Tested Shepherds; Names; CF-Titer
Herd No. 146 aborted	33/71.73%	1:8/1:16–24.24% 1:32/1:64–45.45% 1:128/1:256–30.30%	1. D. I.–1:64 2. P. G.–1:128
Herd No. 2 (39 sheep): 11 cases of abortions, stillbirths, and births of non-viable lambs; 5 retained placentas.	16/41.02%	1:8/1:16–37.5% 1:32/1:64–37.5% 1:128/1:512–25%	1. K.B.L.–1:8 2. P.M.M.–1:64 3. R.S.R.–1:128
Herd No. 3 (31 ewes): 14 cases of abortions, premature births and stillbirths; 6 retained placentas.	15/48.38%	1:8/1:16–26.66% 1:32/1:64–46.46% 1:128/1:256–26.66%	4. D.B.R.–1:64 5. M.S.CH.–1:32 6. –1:128

Of interest are identified by us cases of infection with Coxiella burnetii in patients with lymphadenopathy and neurological symptoms, an initial contingent of clinics in hematology and neurology [611]. The sick TS.G is a practicing veterinary surgeon with generalized acute Q fever accompanied by massive involvement of many groups of lymph nodes (mediastinal, inguinal, etc.). The patient also has epileptic-like seizures. He performed clinical activity with ruminant animals from farms found with Q fever. Etiological diagnosis is made serologically by double examination and the establishment of seroconversion (I serum sample—1:64, II—1:256) and culturally by isolation of C. burnetii from cerebrospinal fluid. Targeted treatment with tetracycline antibiotics led to full clinical recovery and the gradual reduction and clearance of the serological titers. In the differential diagnosis were rejected suspicions of Hodgkin's disease and epilepsy. Serological tests for chlamydiosis, toxoplasmosis, brucellosis, and echinococcosis were negative. Similar positive response was achieved in patients Z.V. with lymphadenopathy and serologically proven infection with C. burnetii [597, 611].

Q fever was observed in sick N.V., laboratory veterinary surgeon performing daily serological and virological work with clinicopathological and passage specimens: blood, aborted fetuses, and placentas from ruminant animals, secretions and excretions and chick embryos, and cell cultures infected with C. burnetii. The disease had a protracted development and was accompanied by fever, fatigue, pneumonia, and myocardiopathy. Serological response is characterized by the accumulation of CF antibodies against C. burnetii with titers reaching 1:256 [597].

In D.K.O., veterinary surgeon—virologist—infection is implemented in air-dust path during a visit to a farm with epizootic of Q fever in goats. Develops generalized disease with an enlarged liver (hepatitis) and joint involvement. In blood serum significant titers of specific antibodies and diagnostic seroconversion were revealed [597].

8.24.2 Atypical Pneumonias

In Table 8.16 are presented summarized results of studies on etiological role of C. burnetii in atypical pneumonias in humans for the period 1993–2000 g [599, 608] From surveyed by MIFT and CFT total of 14353 patients were seropositive 2150 (14.97 %). Much of the persons concerned are not professionally engaged in animal husbandry and processing of raw materials from it. For the years, the proportion of pneumonias associated with Q fever varies from 9.39% for 1998 to 28.53% for 1993, when the epidemic

Table 8.16 Serological testing for Q fever of sick people with pneumonia for the period 1993–2000

Year	I	II	III	IV	V	VI	VII	VIII	IX	X	XI	XII	Total
1993	70/8[a]	85/3	92/6	85/9	34/7	337/130	482/152	158/64	137/15	110/6	77/4	89/14	1756
	11.42%[b]	3.52%	6.52%	10.58%	20.58%	38.57%	31.53%	40.50%	10.94%	5.45%	5.19%	15.73%	418
	1.91%[c]	0.73%	1.43%	2.15%	1.67%	31.12%	36.36%	15.31%	9.58%	1.43%	0.96%	3.34%	28.83%
1994	91/5	108/24	105/20	132/20	127/11	212/31	95/13	78/10	93/13	119/10	153/10	108/7	1421
	5.49%	22.22%	19.04%	15.15%	8.66%	14.62%	13.68%	12.82%	13.97%	8.40%	6.53%	6.48%	174
	2.98%	13.79%	11.49%	11.49%	6.33%	17.83%	7.47%	5.74%	7.47%	5.71%	5.74%	4.03%	14.24%
1995	110/12	132/15	138/25	161/17	226/43	192/42	234/34	82/3	99/9	182/11	213/35	180/12	1967
	10.90%	11.36%	18.11%	10.55%	14.02%	21.87%	14.52%	3.65%	9.09%	6.04%	12.98%	6.66%	258
	4.66%	5.84%	9.68%	6.58%	16.66%	16.27%	13.19%	1.16%	3.48%	4.26%	13.57%	4.65%	12.48%
1996	185/25	331/47	266/45	314/84	389/81	316/98	217/57	80/8	126/16	192/24	199/36	173/29	2788
	13.51%	14.14%	16.91%	26.75%	20.82%	30.01%	26.26%	10.00%	12.69%	12.50%	18.09%	16.76%	550
	4.54%	8.54%	8.18%	15.27%	14.72%	17.81%	10.36%	1.47%	2.92%	4.36%	6.54%	5.27%	19.72%
1997	153/14	105/5	134/24	122/21	186/56	201/30	152/23	108/23	73/15	121/17	105/14	141/20	1601
	9.15%	4.76%	17.91%	17.21%	30.10%	14.72%	21.71%	21.29%	20.54%	14.04%	13.33%	14.18%	262
	5.14%	1.83%	8.82%	7.72%	20.55%	11.02%	12.14%	8.45%	5.51%	6.25%	5.14%	7.35%	16.36%
1998	92/1	96/9	177/18	196/18	184/27	209/21	179/29	63/9	128/14	212/11	166/6	128/9	1830
	1.08%	9.37%	10.16%	9.18%	14.67%	10.04%	16.20%	14.28%	10.93%	5.18%	3.61%	7.03%	172
	0.58%	5.24%	10.47%	10.47%	15.69%	12.20%	16.86%	5.23%	8.13%	6.39%	3.48%	5.24%	9.39%
1999	181/16	200/9	173/8	196/14	133/19	146/14	133/19	78/15	94/17	122/19	128/14	85/11	1669
	8.83%	4.50%	4.62%	7.14%	14.28%	9.58%	14.28%	19.23%	18.08%	15.57%	10.93%	12.44%	175
	9.14%	5.14%	4.57%	8.00%	10.86%	8.00%	10.86%	8.57%	9.72%	10.86%	8.00%	6.28%	10.48%
2000	142/14	148/15	157/20	165/19	135/13	135/20	77/14	36/8	118/12	88/11	54/3	66/2	1321
	9.85%	10.13%	12.73%	11.51%	9.62%	14.81%	18.18%	22.22%	10.16%	12.64%	5.55%	3.03%	151
	9.28%	9.93%	13.25%	12.58%	8.60%	13.25%	9.27%	5.29%	7.94%	7.29%	1.98%	1.32%	11.43%
TOTAL	1024/	1205/	1242/	1371/	1414/	1748/	1569/	683/	866/	1146/	1113/	970/	14535/
	95	127	166	202	257	386	341	130	111	109	122	104	2150
	9.27%	10.53%	13.36%	14.73%	18.17%	22.08%	21.73%	19.03%	12.78%	9.51%	10.96%	10.72%	14.97%
	4.41%	5.90%	7.73%	9.39%	11.95%	17.95%	15.86%	6.05%	5.17%	5.06%	5.68%	4.83%	

[a] Numerator—number of analyzed samples; denominator—reacted positively;
[b] percentage of positive responses;
[c] relative share of reacting positively to the total number of positive findings.

broke out in Panagyurishte. From the pooled data for the entire period, it shows that diseases occur year-round. In January–March, they move in the range of 4–7% and in April—9.39%, and in the period May, June, and July, figures were 11.95%, 17.95%, and 15.86%.

For the period of four years (2001–2004), serologically confirmed cases of pneumonia caused by C. burnetii are 937 (17.99%) of the surveyed 5207 (Table 8.17). In years, the share of these diseases is correspondingly 12.87%, 20.51%, 9.21%, and 24.22%. The increase in incidence in 2002 and 2004 is due to two outbreaks in Etropole and Botevgrad [597, 607, 609]. And in this period, diseases occur throughout the year: January to March (0–17%); April to August (9.89 to 73.9%); September (6.41 to 24.6%); October (10.15 to 24.13%); November (9.72 to 12.5%); December (from 7.5 to 17.6%).

Comparing the two periods studied 1993–2000 and the 2001–2004 showed increase in the incidence of pneumonia due to Q fever in second with 3.02 percent.

Vascular Diseases

For clarification, the etiological role of C. burnetii in Buerger's disease and other vascular diseases was examined in 280 patients from the Department of Vascular Surgery and Angiology at the Military Medical Academy—Sofia, and the National Center for Cardiovascular Diseases—Sofia, suffering from thromboangiitis obliterans, thrombosis of femoral artery, atherosclerosis of the lower limbs and the like, and a control group of 289 patients with varices and angioneurosis [599, 608].

In the study by MIFT and CFT of 234 patients with thromboangiitis obliterans were found antibodies to C. burnetii phase II at 34.61% and to phase I in 16.23%.

In the remaining 9 tested with thrombosis of the femoral artery, positive were 44.44% to II phase and 33.33% to I phase. Of the 10 patients with atherosclerosis of the lower limbs, 30% were seropositive (II phase) and 10% (phase I); out of 6 patients with the condition after bypass, positive were, respectively, 66.66% and 16.66%; of 21 with superficial thrombophlebitis, serologically positive were 14.28% (phase II) and 9.52% (phase I).

In the control group of 289 patients, antibodies to C. burnetii phase II in low titers were detected in 3.8%, and against phase I lacked.

Conducted etiological treatment led to improved clinical picture and decrease in serological titers. The results demonstrate the involvement of C. burnetii in the etiology of Buerger's disease and other diseases with acute or chronic course [599, 608].

Table 8.17 Serological testing for Q fever of people with pneumonias in 2001–2004

2001				
Month	*Tested/+ (No./%)*		*Month*	*Tested/+ (No./%)*
January	62/0		July	53/3 (5.66%)
February	56/6 (10.71%)		August	48/10 (20.83%)
March	99.9 (9.09%)		September	91/7 (7.69%)
April	47/12 (25.53%)		October	29/7 (24.13%)
May	53/11 (20.75%)		November	32/4 (12.50%)
June	746/9 (19.56%)		December	17/3 (17.64%)
		Total 633/81 (12.87%)		
2002				
Month	*Tested/+ (No./%)*		*Month*	*Tested/+ (No./%)*
January	119/17 (14.28%)		July	126/34 (26.98%)
February	107/9 (8.41%)		August	119/88 (73.94%)
March	107/19 (17.75%)		September	126/15 (11.90%)
April	117/24 (20.51%)		October	128/13 (10.15%)
May	429/168 (35.25%)		November	168/18 (10.71%)
June	369/98 (26.55%)		December	113/20 (17.69%)
		Total 2148/448 (20.51%)		
2003				
Month	*Tested/+ (No./%)*		*Month*	*Tested/+ (No./%)*
January	131/13 (9.92%)		July	90/5 (5.55%)
February	98/9 (8.16%)		August	42/6 (14.28%)
March	196/10 (5.10%)		September	78/5 (6.41%)
April	118/13 (11.01%)		October	92/11 (11.95%)
May	82/14 (17.07%)		November	72/7 (9.72%)
June	78/10 (12.82%)		December	80/6 (7.50%)
		Total 1139/105 (9.21%)		
2004 (to September Inclusive)				
Month	*Tested/+ (No./%)*		*Month*	*Tested/+ (No./%)*
January	107/8 (7.47%)		July	105/32 (30.47%)
February	76/9 (11.84%)		August	109/18 (16.52%)
March	113/14 (12.84%)		September	97/21 (21.64%)
April	91/9 (9.89%)		–	–
May	330/89 (26.96%)		–	–
June	380/100 (26.31%)		–	–
		Total 1251/303 (24.22%)		
		Total for 2001–2004, Tested 5207/ Positive 937 (17.99%)		

8.24.3 Chronic Endocarditis

Using MIFT and CFT were obtained serological evidence of infection by Coxiella burnetii in 22 cases of chronic endocarditis [599, 608]. From the heart valves of six operated patients were isolated 6 strains C. burnetti—4 after inoculating 7-day chicken embryos immunosuppressed with gamma rays (3.5–4.5 G) and 2 after inoculation of young white mice. The identification of strains is achieved by immunofluorescence method.

Chronic Q fever manifested primarily as endocarditis is severe and often ends in death. Data obtained by us are grounds for advanced purposeful search for etiology associated with Coxiella burnetii in this cardiac pathology.

8.24.4 Analysis of the Serological Data

Above were presented results from made numerous serological examinations for Q fever of blood samples from 8 species of domestic mammals, 4 species of poultry, 13 species of wild mammals, four species of wild birds and from people. Received by us data are based on surveys of more than 305,000 serum samples conducted using CFT and micro-IF test.

In earlier studies of the problem in Bulgaria by CFT were collected epidemiological data on infection caused by C. burnetii, mainly in sheep and goats [5, 6, 16, 17, 43, 47, 49, 57]. Extremely large volume of our studies confirm the effectiveness of this serological method. World Organisation for Animal Health—OIE [381]—presented updated standards for veterinary laboratory diagnostics and reiterated that the three most suitable and commonly used serological tests with Q fever are CFT, MIFT, and ELISA.

The complement fixation test is widely used in a large number of laboratories in many countries [381, 409]. A key feature of the CFT is its high specificity. Proving antibodies against antigens of C. burnetii in phases I and II, which involves registering and tracking seroconversion and dynamics of antibodies, allows the differentiation of acute and chronic infections [118, 217, 395, 397]. Established persistence of complement fixing antibodies also make possible the retrospective proof of the infection [448, 458]. In the reaction for serological testing of animals is most commonly used antigen in the second phase (Bouvery et al. 2003 [105]). Recent studies of Q fever in humans in the UK are also based on CFT with second-phase antigens—Van Woerden et al. [533]. Investigations of people in Oman under the same methodological conditions was carried out in the study of Scrimgeour et al. 2005 [467]. Dolće et al. 2003 [167] also used CFT with a second-phase antigens for serological examination of sheep in Quebec, Canada. Parallel titrations of KC antibodies in the blood serum of animals and humans in the first and second phases are useful for tracking the dynamics of epizootic process in foci Q fever and to assess the freshness of the disease [31, 126, 284, 510]. Russo [448] and Sanchis et al. [458] noted that CFT gives excellent results in the routine diagnosis of Q fever in the herd, especially in outbreaks of abortion and related conditions. We support the above opinions of experts in the field of veterinary and human medicine.

The method indirect immunofluorescence (IFA, MIFT) has much greater use in human medicine [119, 167, 195, 355, 381, 518]. Against antigens of C. burnetii (phases I and II) are defined antibodies in immunoglobulin fractions IgG, IgM, and IgA, which allows assessment of the development of the infectious process and the state of immunity of the body and the differentiation of acute and chronic Q fever in humans. At this stage, however, in domestic animals, antibody response against phases I and II established in MIFT and immunoglobulin classes are not well understood [381], which defines the lesser application of the method in veterinary medicine.

MIFT was widely used for serological tests on different continents' sick people—mostly with atypical pneumonia, endocarditis, and vascular diseases—as well as for sero-epidemiological studies. Both serological reactions—MIFT and CFT conducted with antigens in phases I and II—are recommended by the WHO [4, 126, 225, 273, 343, 355, 510]. The application of both tests on a large number of serum samples enables us to uncover many cases of infection with C. burnetii in humans during epidemics and smaller outbreaks and in sporadic disease occurring throughout the year inbetween epidemic periods. The method MIFT was also applied in investigations of cows from farms with high rates of abortions, detentions of placentas, and endometritis. Parallel titration of antibodies in CFT showed some excess of this test to the detection of seropositive reagents—80.95% against 71.42% (MIFT). Match results of both tests was 52.38%, and a significant part of serologically positive animals are found alone in one or the other method: 28.57% (CFT) and 19.04% (MIFT). The results obtained show that MIFT may be part of the methods for serological diagnosis of Q fever in cattle. Relatively light staging of the reaction and the visibility of results are also positive moment. Nevertheless, we believe that now using MIFT alone in animals obtain more data on the serological response to the agent in both phases and immunoglobulin structure of antibodies, thereby supporting the above-cited vision of OIE [381]; especially in cows with that clinical characteristics, safer would be the results of co-administration of MIFT and CFT as obvious correlation between the two methods is not constant and probably depending on the phase of infection with Coxiella burnetii.

Statistical treatment of results of our serological examination at Q fever in cattle, sheep, and goats, confirmed by the direct methods for detecting the agent and its isolation, provides new evidence for the wide dissemination of this infection in Bulgaria among these species, currency, economic, and zoonotic importance of which is recognized by many contemporary authors [106, 167, 232, 264, 315, 351, 385, 577]. Systematic sero-epidemiological studies throughout the country enabled us to build a comprehensive view of the state of the problem Q fever in animals and related diseases in humans. For the last five years of the study (2002–2006), Q fever in cattle was detected in 172 settlements from various districts and regions. In sheep and goats, number of detected foci is correspondingly 105 and 70. In connection with these facts, we believe it is imperative the continuation of the annual serological screening of herds of domestic ruminants, both in clinical and epidemiological evidence for possible detection of new outbreaks as well for tracking the dynamics of old foci Q fever in the country. Serological studies are needed and in the quarantined newly purchased from abroad animals before placing them on farms. Until 1997, these studies were mandatory in the preceding 20–25 years. The fact that among imported animals in quarantine were discovered seropositive animals certainly raises the question of the restoration of the previous useful practice and the introduction of strict border control of Q fever. The need for prophylactic serological tests for the disease is particularly appreciable before the movement of ruminants in the country in connection with trade, formation of new herds, or completion operation of existing ones. In this large study, we have repeatedly established themselves outbreaks of Q fever in connection with the uncontrolled movement and trade of sheep, goats, and cattle. Similar observations have previous explorers of Q fever in Bulgaria: Ognianov [43]; Pandarov [47, 49]; Genchev et al. [19]. Serological studies with complex of other events will contribute to limiting the disease in the country.

The establishment of serological status of the C. burnetii infection among domestic ruminants is a compulsory element in studies to clarify the source of infection in the outbreaks of Q fever in humans. Convincing examples of this are the outbreaks in Botevgrad/2004/and Etropole/2002/. Serological data obtained clearly showed the connection between the outbreaks in humans and agricultural foci Q fever in the studied areas. For the common etiology of the disease in animals and humans witness and performed our purposeful serological examinations of goats, sheep, and cows owned by clinically ill persons from atypical pneumonia, where Q fever has been proven etiologically. Significant methodological detail in clarifying the serologic status of ruminants in connection with an epidemic of Q fever among humans is receiving serological data for animals in adjacent areas to directly affected area, where there is no disease in humans and also from other areas of the country. These data clearly highlight the trend of high seropositivity in the focus of epizootic among animals, where epidemic is developing among the human population, while in neighboring areas it has moderate seropositivity and in other regions of the country reveals relatively low seropositivity. Botevgrad, 2004, is a typical pattern for an established trend with a range of serologically positive for C. burnetii goats—66.33%, sheep—46.66%, cattle—33%; in neighboring regions of Botevgrad—2.6 times less coverage; in other parts of the country—5,2 times lower seropositivity. Very significant is the comparative analysis of serological data in Botevgrad in 2004, when the epidemic broke out in humans with those of 2002/Etropole/when there are no human diseases in Botevgrad, but the city is adjacent to Etropole—agricultural focus of Q fever and center of epidemic among the population. Snowballing of emergence and increase in diseases in humans (2004) corresponded to a significant increase in the percentage of the serologically positive animals— 3 times in cattle, 2 and a half times in goats, and more than 2 times in sheep.

In the complex of etiologic research on diseases causing major economic damage—abortions, premature births, stillbirths, or non-viable offspring— serological reactions take significant share. Commonly studied groups of at least five to ten animals in a herd having these problems. A similar recommendation makes OIE [381, 409]. Self-serological diagnosis is possible in the investigation of single and two-time blood samples. They determined individual KC titer and geometric mean titers. In the statistical analysis and interpretation of the results, an important place has data obtained by titration of all sera taking into account the period of receipt of the blood sample after an abortion or birth of dead or non-viable offspring. The object of analysis were other indicators: the percentage of positively reacted for C. burnetii animals having pathology of pregnancy (abortions, etc.) in the herd or farm during the first 30 days after abortion; presence and percentage of seropositive reagents in later periods; the existence, scope, or absence of infection with the agent in control groups—ruminants with normal births in infected and uninfected farms; pregnant animals; healthy non-pregnant animals.

The study of twofold serum samples in cases of abortion and related conditions in order to assess the dynamics of the antibody is effective. The establishment of diagnostic seroconversion undoubtedly indicated the etiology in some of these clinicalforms of Q fever. The disclosure in a number of farms on abortion and related conditions of the pathology of pregnancy with dominant or partial involvement of C. burnetii confirmed and enriched the known facts about the etiological role of C. burnetii in some of these diseases in domestic ruminants [43, 44, 47, 228, 275, 336, 386, 457, 531, 573].

Very informative are results of the experimental infections that we implemented in sheep and cows. The received serological evidence points to the accumulation of the second-phase and the first-phase antibodies against C. burnetii with pronounced dynamics. There were significant differences in terms of the occurrence of two types of antibodies, titer values, and length of detention. We confirmed the findings of other authors [31, 126, 284, 397, 510] for early positivation of CFT with antigens in phase II, which explains their more wide use for diagnostic purposes.

We find interesting and significant research results on the duration of storage of CF antibodies in blood sera of sheep infected spontaneously of active foci of Q fever. The observations covering 5-year period show a long persistence of the antibodies in the second phase—up to 3, 4, and 5 years of low titer levels in progressively decreasing number of animals. For the first-phase antibody is characteristic not just by their later occurrence in the blood serum, but also their relatively shorter retention—up to 1.5 to 2 years at low values. The data obtained can be used in attempts to retrospectively demonstrate the disease in sheep.

A detailed review of the literature indicates that the rams, unlike the sheep, have not been subject to targeted etiologic studies on Q fever. This is understandable because sheep often found clinical pathology with mass character, leading to economic losses. On the other hand, it is recognized the great epizootiological and epidemiological importance of this kind as a source of infection for other animals and humans at parturition and lactation. In addition, the lack of data on clinical course of the disease and the possible transmission of C. burnetii sexually contribute to low interest to rams in the studies on Q fever. We have taken great volume study involving 1662 animals from 30 farms and two quarantine facilities. In 18 of the farms had been previously established Q fever in sheep including abortions and unviable newborns in 6 of them. The remaining farms were free of Q fever or were of unknown status. The aggregated results showed that 56 rams (3.36%) of 15 farms were serologically positive for C. burnetii. Without exception, positive seroreagents belonged to farms with proven Q fever in flocks that occurs among sheep with clinical manifestations or latent. This means that the rams are part of the epidemiological chain and most likely involved in the spread of infection by the pathogen transmission with feces and urine.

The smaller percentage of seropositive rams compared with sheep, apparently, is associated with greater resistance to male sex in sheep populations against infection with C. burnetii.

The articles of Ognyanov et al. [45, 43] and Hore [249] described the Q fever in lambs, manifested by mass respiratory diseases and pneumonia in adult goats in herds where both were observed abortions [43]. Overall, however, studies in this area are currently limited. We turned our attention to a number of places where there were clinical and epidemiological indications of such diseases and expanded research on other affected age groups in small ruminants and for the first time studied calves with pneumonia. Serological results obtained conclusively confirm the nature of pneumonia in lambs of 1–6 months of age and also reveal the involvement of C. burnetii in the etiology of pulmonary inflammation in lambs (12–14 months) and adult sheep. In addition to studies of pneumonia in sheep and lambs as a result of natural infection with C. burnetii, we have reproduced the disease experimentally by intratracheal or intranasal infection of lambs and sheep. As a result, developed general illness with clinical picture of pneumonia of interstitial type and expressive serological response with a specific dynamics were tracked for 120 days. Similar positive findings were obtained in spontaneously infected goats from different age groups of both sexes with clinical signs of acute bronchitis or interstitial pneumonia. Registered individual CF-titers, GMTs, and the scope of infection have shown high susceptibility of goats to infection by the inhalation route with C. burnetii resulting in developed respiratory symptoms and expressed serological response; especially the first obtained serological evidence of infection with C. burnetii in calves aged 2–4 months has to be mentioned. These results extend the etiologic spectrum of pulmonary inflammation in the type and together with the above findings highlight even more the economic significance of Q fever. All three groups' serology in respiratory diseases were confirmed through virological isolation of the pathogen from the lungs of affected small ruminants and cows.

Applying the approach of the complex etiological diagnostics including serology, direct EM, and isolation of the causative organism, proved severe acute mastitis in sheep caused by C. burnetii occurring as a distinct clinical form of mass character. In the studied mastitis outbreaks, representing first description in the literature were regularly accumulated specific antibodies. In titration of blood sera in appropriate terms were established predominantly high CF-titers. Antibodies against C. burnetii also were found in blood samples from cows with acute, subacute, and subclinical mastitis. Of milk secretions of serologically positive cows with acute mastitis was isolated C. burnetii.

To prove the role of C. burnetii as the cause of mastitis in cows can be successfully used other approaches—detection of IgG antibodies (ELISA) in the milk serum [385] or demonstration of the antigen of the pathogen in the milk [151].

Our serological and virological investigations and direct methods of indicating the agent confirmed the spread of Q fever in other economically significant disease conditions in cows—endometritis, metritis, infertility, prolonged reduction in milk secretion due to oligosymptomatic or latent inapparent infection by C. burnetii. The effectiveness of serological examinations is evident from the disclosure of the latent infection with C. burnetii in bulls from breeding centers that do not show clinical signs. These data are important for the timely identification of the infected animals and taking appropriate measures. The fact that seropositivity for C. burnetii bulls is found in selection centers raises the question of mandatory periodic serological testing of animals. This will prevent the risk of turning the affected centers in foci of Q fever.

In other farm animals—*buffaloes, pigs, and horses*—seroepidemiological data show a slight or insignificant spread of infection with C. burnetii. This information is also based on extensive studies that have covered a large number of animals (buffalo 1180, 2514 pigs, 1284 horses) from different regions of the country. The seropositive buffaloes (1.43%) were asymptomatic. They came from three farms in three areas. In seven other farms lacked positive reactions. Therefore, in some herds buffaloes there is poor circulation of the agent, while in others it is missing. Among other possible factors explaining this fact can be found in the largest natural resistance of buffalo populations to infection with C. burnetti. In the late fifties of the twentieth century, Zarnea et al. [577] in Romania and Shindarov et al. [66] in Bulgaria reported for significant percentage of serologically positive horses in active agricultural foci of Q fever. Approximately in the same period Nikolov [40] investigated serologically 520 horses and found positive reactions in 12 cases (2.3%). Regardless of the contradictory data in that early stage of knowledge of the problem, it can be assumed that the then economic conditions in which the largest number of horses were used as a living force in agriculture and their pasture breeding are a prerequisite for more widespread infection. In our investigations were established small percentage (0.66%) of specific antibodies in horses in it for an extended period of time (1989–2004). The analysis of our own data from previous periods—1984–1988 years (0.89%) and 1977–1983 years (1.02%)—shows the long-term trend for rare detecting just single seropositive horses—very poor circulation of C. burnetii in that species. The situation is similar in pigs. In early studies, Angelov et al. [8, 9] and Gardevska and Panaiotova [25] investigated, respectively, 83 and 100 sows, but antibodies against C. burnetii not been established. Nikolov [40] investigated 336 pigs and 8 of them (2.38%) were serologically positive. The same percentage established Gaon and Gall [204] in pigs in Yugoslavia. Howe et al. [252] did not detect antibodies against C. burnetii in the testing of 396 pigs from different prefectures in Japan. The systematic serological screening, which was conducted from 1977 to 2004, revealed only two seropositive pigs—1984–1988 (1); 1977–1983 (1). In the period 1998–2011 positively reacted pigs are not found. Obviously, at this stage, the involvement of pigs in the formation of agricultural foci of Q fever should be excluded.

Our results of the serological tests confirmed the susceptibility of *dogs* to infection with C. burnetii. The attention of the authors studied the problem was mainly aimed at dogs inhabiting agricultural foci of Q fever assuming that the spread of infection in them is dependent on its presence and intensity in ruminants [40, 314, 354]. Serological and cultural studies in dogs are carried out in connection with suspicions that they may be a source of infection for humans [115, 126, 290, 382, 429]. Discovered is different degree of prevalence of this species to C. burnetii: 28, 8–30, 3% in Africa [323]; 15% in Japan [252]; 13% in Germany [323]; 3.4%–8% in Bulgaria [40, 58]. Conducted by us, serology on dogs from active agricultural foci revealed many seropositive animals (49.45%). Completely new element is the serological testing for C. burnetii of homeless wandering dogs due to an extraordinary increase in their number over the last 10 years. In titration of blood sera from 98 such dogs in Sofia, 23.46% were serologically positive. This range of the infection caused by Coxiella burnetii is significant of the conditions of the capital city and deserves special more detailed studies.

In infected *cats,* zoonotic importance of which was recognized in the last decade, we found an average prevalence of C. burnetii in 37.5%, including 47.05% in stray cats wandering and 26.66% when grown at home. Such parallel also made Japanese researchers–Komiya et al. 2003 [278]. They found specific antibodies in 41.7% in homeless cats and 14.2% when grown at home.

The serological studies of approximately 2,000 poultry from four species revealed the presence of CF antibodies against C. burnetii in hens (4.24%), ducks (5.6%), and geese (4.09%). Serologically positive ducks are found for the first time in the country. The surveyed turkeys were negative. The preceding data for the circulation of agent of such birds in the country date back to the fifties and sixties of the twentieth century: Shindarov et al. [66] reported antibodies in 15% of hens and in 17% of the geese; Gardevska and Panajotova [25] for 8% seropositive turkeys and chickens 2%. Nikolov [40] found no positive reactions in chickens. In the same period, Romanian [578] and Yugoslav authors [204] reported that poultry in their country are free from infection with C. burnetii. Nowadays, the prophylactic serological examinations of poultry should continue as part of the overall epizootic supervision of Q fever and for the timely detection of positive cases that may serve as additional reservoirs and sources of the agent of Q fever.

Our serological investigations of people covered four groups of patients: those working in animal husbandry and veterinary medicine; atypical pneumonias; vascular diseases; chronic endocarditis. Received conclusive evidence of Q fever in the first group confirmed the importance of the disease among professions whose nature requires daily direct contact with the animal reservoir and source of infection. Too valuable in this respect are presented well-documented cases of Q fever in shepherds and sheep bred from them and practitioners and laboratory veterinarians. These results reiterate the importance of the serological tests for Q fever carried out preventive or at the clinical-epidemiological indications among these and other risk professional contingents. Our conclusion on this issue coincides with the opinions of other authors set out in a number of publications [123, 228, 241, 306, 343, 378, 433, 509].

An essential part of our research in humans are the serological investigations of patients affected by pneumonias and must be emphasized that the majority of them did not belong to professional risk groups. To clarify the etiological role of C. burnetii covered the period from 12 years (1993–2004) with almost 20,000 tested sera. For the entire 12-year period, the average total seropositivity at the pneumonia, mostly with atypical course was 15.78%, and for the last four years (2001–2004) was 17.99%. These results clearly indicate the etiological share of the C. burnetii at the atypical pneumonias in humans. Literary data also show that cases of pneumonias caused by C. burnetii are no exception and for them must be thinking targeted, especially considering the region, respectively, the status of domestic animals in terms of Q fever [326, 330, 343, 426]. Undoubtedly, the establishment of Q fever in animals in an area is the basis for focused research in people with pneumonias and other clinical manifestations—fevers, hepatitis, etc. Given the changes in classical epidemiology of Q fever and particularly the deletion of the typical risk in the past and age element, it is appropriate that each open pneumonia should be tested for infection with C. burnetii. Observations show that because of diagnostic omissions some cases of pneumonia associated with C. burnetii are not disclosed, especially for children [126, 265, 273, 313, 449]. The serological methods used in this study provide fast and accurate etiological diagnosis. Obviously, that it can not rely only on clinical examination and epidemiological information. Similar opinion was expressed by Fournier et al. [198]. Experience shows that timely detection of pneumonia caused by C. burnetii and directed etiological treatment are crucial for speed and shortening the duration of therapy, which is undeniable socioeconomic impact.

Subject to our serological examinations were the vascular infections with C. burnetii, also studied by other teams of authors [112, 179, 188, 196, 343, 424]. Specific antibodies detected in thromboangiitis obliterans, thrombosis at the femoral artery, atherosclerosis of the lower limbs, after bypass condition, and superficial thrombophlebitis. The results testify to the involvement of C. burnetii in the etiology of Buerger's disease and other vascular diseases with acute or chronic development, which requires the advisability of such studies to confirm or at least to reject the etiology associated with the pathogen of Q fever. French researchers [196] emphasize that the incidence of vascular C. burnetii infection is probably higher than recorded, making it necessary serology for Q fever, which become routine in all cases of unexplained febrile illness, pain, or weight loss at persons affected by the disease.

Achieving the etiological diagnosis of chronic endocarditis resulting from Q fever confirms the relevance of the problem in this direction. In comparison with the data of Raoult et al. [426], the proportion of endocarditis caused by C. burnetii compared with the total incidence of Q fever in Bulgaria is small, but their number is constantly growing. The need for directed search of possible etiology associated with C. burnetii of these serious diseases is evident. The diagnosis of Q fever at the specified cardiac pathology and implementation of definitive treatment are essential for the prognosis of these disease states.

9

Polymerase Chain Reaction

Mallavia et al. 1990 [311] exhibited the principles of the strategy for detection and differentiation of C. burnetii strains by using the polymerase chain reaction (PCR). The application of PCR, using specific oligonucleotide primers and Taq DNA polymerase to synthesize increasing amounts of DNA from a single template [451], represents a new approach in the diagnosis and studies of C. burnetii and Q fever [489]. The method is applicable to the examination of tissue samples from clinical specimens and strains of C. burnetii propagated in chicken embryos, cell cultures, or laboratory animals. Use fresh, frozen, or paraffin-embedded even materials—placenta, fetal parenchymal organs, bone marrow, liver, feces, heart valves, milk, uterine and vaginal exudates, cerebrospinal fluid, and the like. [380, 488, 489, 554, 605]. The reaction is suitable for a retrospective diagnosis by examining the stored lyophilized and frozen cultures and tissues. According to Fritz et al. [203], PCR techniques permit quantitative analysis of C. burnetii in tissue samples.

It has been found that it is possible amplification of some genes of C. burnetii—htpAB, sodB, gltA, and 16SrDNA [248, 554]. Willems et al. [554] reported the transposon PCR with primers derived from htpAB-related transposon-like sequences which, depending on the isolate, have a minimum frequency of 19 for the genome of C. burnetii. The authors use the method of detection of C. burnetii in cow's milk. Because of the small amounts of cells of the agent in the milk—required special preparation—the concentration of the genomic DNA by a factor of 200 and precipitation with cetyltrimethylammonium bromide. Thus, the method becomes highly sensitive and specific as even a single cell of C. burnetii can be detected in 1 ml of milk [554]. Stein et al. [493] amplified ITS (internal transcribed spacer) of 16S-23S ribosomal DNA (rDNA) of 22 isolates of C. burnetii (from humans, cattle, cats, and ticks) in the USA, Canada, France, Italy, Poland, and the Central African Republic and perform sequencing with automatic laser fluorescent DNA sequencer.

ITS measured 497 base pairs (bp) and encodes isoleucine-tFK and alanine-tPHK. When comparing, the consecutive chain links of the strains disclosed are high levels of similarity (>99%), regardless of the different geographical origins of the strains and their different phenotypic characteristics. It is concluded that the sequencing of 16S-23S rDNA ITS of C. burnetii can be used for the identification of the agent, but is not applicable to studies of epidemiology, virulence, and taxonomy [493].

Some research teams used successfully nested PCR for demonstrating C. burnetii. In Japan, To et al. [522] examine infection in birds and identify the pathogen in 17 (41%) of the 41 surveyed poultry and at the 37 (22%) of 167 wild birds. Greek authors [486] applied the method to confirm positive results in the inoculation of blood from sick people (Crete) in Vero cell cultures and HEL cells, as well as directly in the blood coagulum.

Masala et al. 2004 [336] did extensive research on abortions in Q fever in sheep and goats in Sardinia using serological methods and PCR. From flocks with serological evidence of Q fever in the PCR were studied placentas and fetuses. Although DNA of C. burnetii was amplified from different tissue types, the highest percentage of detection was established in placentas [336].

There are attempts to prove the identification of C. burnetii in ticks [202, 493, 499]. Fritz [202] adapted PCR for quantitative analysis but finds no agent in 1304 copies Ixodes ricinus, collected in the region of Hesse, Germany. Later, also in Germany, Baden-Wuerttemberg Sting et al., 2004 [499] explore in the PCR 1066 ticks Dermacentor (including D. marginatus) of 23 sheep flocks and 49 samples tick excrements from 18 herds. Two positive results are received—one in the tick and the second in the excrements. These results, while confirming the serological data for circulation of C. burnetii in flocks, are not sufficient to conclude on the sensitivity of PCR for the detection of agent in ticks.

Not all clinical materials are suitable for PCR. Blood serums are usually negative or give false-positive results [343]. In this regard, Maurin and Raoult [343] expressed skepticism about the announcement of Japanese authors for the presence of DNA of C. burnetii in 23 (39.6%) of the sera of 58 studied children with atypical pneumonia [521]. As mentioned above [486], isolates from blood in cell cultures and blood clots from human patients can be used for PCR-diagnosis of C. burnetii. In a previous study [369], positive blood cultures from patients with severe (pneumonia) and chronic (endocarditis) Q fever are proven in PCR. As an essential condition for the application of the reaction in these cases, patients have not been treated with antibiotics prior to blood sampling [369].

Yanase et al. [569] reported the detection of the agent by PCR in 5 of the 10 test samples of dust from dairy farm in Japan. Van Woerden et al., 2004 [533] reported unsuccessful attempts for detecting of C. burnetii by PCR in samples of straw and dust from the factory premises of the processing plant in Britain, where has erupted etiologically proven epidemic of Q fever. Since contaminated with C. burnetii straw, hay, and dust that are considered as possible sources of infection in sporadic cases and outbreaks [136, 450, 452, 432], the reasons for the negative results can be found in several directions—insensitivity of the method with respect to similar samples, technical problems with the placement of the reaction, or simply in the absence of the microorganism in the tested materials.

9.1 Detection, Identification, and Differentiation of Coxiella burnetii by Polymerase Chain Reaction

Below are presented data on the use of PCR's three techniques: normal (conventional) PCR; REP-PCR; RAPD-PCR with primers, production of ISOGEN (Netherlands) (Table 9.1) [612, 613] and the appropriate laboratory procedures for receipt of the DNA from microorganisms [266, 383, 401, 489], the amplification of DNA [183, 184, 535, 551], detection of amplification product [453], and computer account of the matrix DNA profiles (fingerprints) [80, 187, 376, 480].

Table 9.1 Sequence of the oligonucleotide primers used in the polymerase chain reaction

Primers		Sequence (5′ μ 3′)
Conventional PCR		
Coxiella	CB–1	ACT CAA CGT ACT GGA ACC GC
burnetii	CB–2	TAG CTG AAG CCA ATT CGC C
Rickettsia	Rr 190.70p	ATGGCGAATATTTCTCCAAAA
	Rr 190.701n	GTTCCGTTAATGGCAGCATCT
Chlamydia	CTU	ATGAAAAAACTCTTGAAATCGG
	CTL	CAAGATTTTCTAGAYTTCATYTTGTT
Rep-PCR		
REP 1R-I		III ICGICG ICA TCI GGC
REP 2–I		ICG ICT TAT CIG GCC TAC
RAPD-PCR		
P5		CGGCCC CGG T
OPL 04		GAC TGC ACA C

9.1.1 Conventional PCR

The results of these experiments are presented in Figure 9.1.

In conventional PCR for detecting of C. burnetii, we used the pair of primers CB-1 and CB-2, and for chlamydia, primers CTU and CTL. From the results of Figure 9.1 (B), it is seen that the C. burnetii-resulting amplification products are with a size of 260 bp (base pairs). Upon chlamydia strains (A)DNA the products are of size 900 bp. Normal PCR with DNA from strains of Rickettsia spp was used pair of primers 190.70r Rr and Rr 190.701n, sequenced by gene rOmpA. In amplification with these primers were synthesized DNA products with a size 590 bp from both genomic DNA of R. rickettsii, and by R. conori and R. sibirica (Figure 9.1C).

The results obtained show that the normal (conventional) PCR is a suitable method for differentiation to a level of species of microorganism, but it is not suitable for species differentiation within representatives from a given genus—Coxiella, Rickettsia, and Chlamydia. For this purpose, we adopted a different methodological approach, namely the use of other specific PCR techniques described below [612, 613].

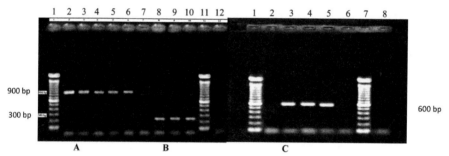

Figure 9.1 Amplification products from conventional PCR with DNA from strains Coxiella, Chlamydia, and Rickettsia spp. **Legend: Lines 1 and 11**—DNA marker 100 bp; **A. Lines 2–6**—900 bp amplificates from: 2. Chlamydia trachomatis M (urethritis); 3. C. psittaci ornithosis (canary); 4. C. psittaci CP3 (ornithosis pigeon); 5. C. psittaci GA (abortion in sheep); 6. C. psittaci T-1f (abortion in goats); 7 and 12, negative controls (samples without DNA); **B. Lines 8–10**–260 bp amplificates from: 8. Coxiella burnetii, Henzerling (reference human strain); 9. Coxiella burnetii M-44 (per person); 10. Coxiella burnetii (aborted cow); 1 and 7 DNA marker 100 bp; 2 and 6 negative controls (samples without DNA); **C. Lines 3–5**—amplificates with a size 590 bp of: 3. Rickettsia rickettsii; 4. R. conorii; 5. R. sibirica; 8. blank.

*Indications 900 bp, 600 bp, 300 bp—markers.

9.2 Correlation Analysis of Strains of Coxiella burnetii, Chlamydia, and Brucella, and Representatives of α-1 (Rickettsia) and α-2 Sub-divisions of the Proteobacteria, Based on the REP-PCR

In the study, 36 strains of group according to their origin (genus) in five main groups are used such as Coxiella, Rickettsia, Chlamydia, Brucella, and Proteobacteria. In REP-PCR amplification with primers, REP 1R-I and REP-2-I were the observed DNA profiles consisting of different numbers of fragments of a certain size (Figure 9.2).

In four of five surveyed strains Chlamydia (lines 1–4 and 36), the number of fragments is between 9 and 23 as two are common to them all. In each species, one or more specific fragments shared with some of the other species which determines the genetic relationship between them are observed. In examining the profiles of Chlamydia trachomatis M and C. trachomatis W (lanes 1 and 2), it is seen that they are composed of 23 fragments as 14 of them have the same size. The strains Chlamydia psittaci (lines 3 and 4) hold DNA profiles consist of 17 and 19 fragments of which 5 are of uniform size. C. psittaci ornithosis (line 3) has 12 general fragments with C. trachomatis M and 9 with C. trachomatis W. DNA profile of C. psittaci CP 32 indicates 4 general fragments with C. trachomatis M and 3 with C. trachomatis W. For its part, C. psittaci GA (line 36) has only 2 identical fragments with others C. psittaci and one with C. trachomatis, located on different levels of DNA profiles due to their different sizes.

The profile of the three strains Coxiella burnetii (lines 5–7) is composed of 18–20 fragments, of which 11 are of equal size. This speaks of conservatism in their genome. Coxiella burnetii 4 and 5 have two identical fragments.

In the strains, Rickettsia (lines 32–34) was observed expressed polymorphism—their profile is composed of eight to nine fragments and only one of them is the same size in all three tested strains. Rickettsia rickettsii have one common fragment with Rickettsia coronii and Rickettsia sibirica, located on different levels. Rickettsia coronii and Rickettsia sibirica have two common fragments. Rickettsia (lines 32–34) was observed expressed polymorphism—their profile is composed of eight to nine fragments and only one of them is the same size in all three tested strains. Rickettsia rickettsii have one common fragment with Rickettsia coronii and Rickettsia sibirica, located on different levels. Rickettsia coronii and Rickettsia sibirica have two common fragments.

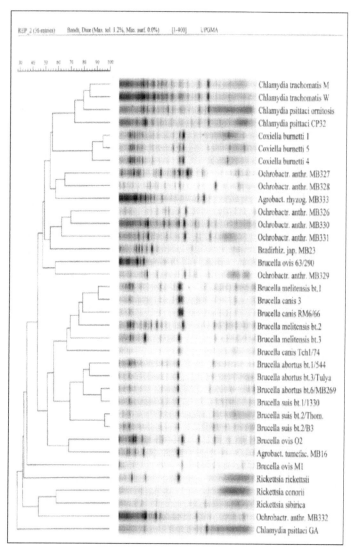

REP_2 (36 entries)   Bands, Dice (Max. tol. 1.2%, Min. surf. 0.0%)   [1-400]   UPGMA

Chlamydia trachomatis M
Chlamydia trachomatis W
Chlamydia psittaci ornitosis
Chlamydia psittaci CP32
Coxiella burnetti 1
Coxiella burnetti 5
Coxiella burnetti 4
Ochrobactr. anthr. MB327
Ochrobactr. anthr. MB328
Agrobact. rhyzog. MB333
Ochrobactr. anthr. MB326
Ochrobactr. anthr. MB330
Ochrobactr. anthr. MB331
Bradirhiz. jap. MB23
Brucella ovis 63/290
Ochrobactr. anthr. MB329
Brucella melitensis bt.1
Brucella canis 3
Brucella canis RM6/66
Brucella melitensis bt.2
Brucella melitensis bt.3
Brucella canis Tch1/74
Brucella abortus bt.1/544
Brucella abortus bt.3/Tulya
Brucella abortus bt.6/MB269
Brucella suis bt.1/1330
Brucella suis bt.2/Thom.
Brucella suis bt.2/B3
Brucella ovis O2
Agrobact. tumefac. MB16
Brucella ovis M1
Rickettsia rickettsii
Rickettsia conorii
Rickettsia sibirica
Ochrobactr. anthr. MB332
Chlamydia psittaci GA

Figure 9.2   Results of clustering analysis of strains of Coxiella, Rickettsia (α-1 sub-division of Proteobacteria), α-2 sub-division of Proteobacteria, Chlamydia, and Brucella, by computer program UPGMA, based on REP-.

The fragmented profile of species of the genus Brucella contains 9–22 fragments, as disclosed in the present study only 14 strains had a general fragment. The analysis by type shows that the group of Brucella abortus has a DNA profile composed of 15–18 fragments, of which 13 are of equal length.

The DNA profiles in Brucella canis contain between 9 and 17 fragments of which 7 are identical in size and in Brucella ovis—9 to 20 fragments as seven are the same. The strains Brucella melitensis have DNA profiles composed of 18–22 fragments of which six are of uniform size. In Brucella suis, the strains have 13–16 fragment and only two are identical. The most conservative profile has Brucella abortus.

In the surveyed ten strains of α-2 sub–division, Proteobacteria differences are observed between 14 and 30 which do not show fragments of the same size; that is, the representatives of this group exhibit the highest polymorphism.

In the present study, the different types of Brucella (lines 15, 17–29, and 31) and bacteria from α-2 sub-division of Proteobacteria (Ochrobactrium, Agrobacterium, and Bradirhiz. Jap.) (Lines 8–14, 16, 30, 35) to which has been shown that REP-PCR generates very similar DNA profiles (Tcherneva et al. [507]) are included for comparison. The analysis of the selected groups of strains shows that the α-2 Proteobacteria genotypically are closest to Rickettsia (lines 32–34), then to Brucella (lines 15, 17–29, 31), and Chlamydia (lines 1–4), and outermost are from the Coxiella.

The DNA profile of C. psittaci GA (line 36) is closer to the α-2 bacteria rather than to the other Chlamydia.

The profile of Coxiella burnetii (lines 5–7) is the most conservative and is characterized at the DNA level from the other groups of strains. Notwithstanding this conservatism, the genetic analysis of the investigated strains Coxiella burnetii indicates the same-size fragments with Proteobacteria, from 1 to 5; with Ochrobactrium (lines 8–9; 11–13; 16 and 35); and by two with Agrobacterium (lines 10 and 30) and Bradirhiz. jap. MV23 (line 14). Coxiella burnetii show five identical fragments with Brucella ovis O2 (line 29); four with Br. abortus, bt. 1, 3 and 6 (lines 23–25); 4 with Br. suis, bt. 1 and 2 (lines 26–28) and two with Br. canis (lines 18, 19 and 22). Coxiella burnetii 1, 4, and 5 have three identical fragments with Rickettsia coronii (line 33), one with R. rickettsii (line 32), R. sibirica (line 34), and Chlamydia trachomatis (lines 1 and 2).

REP-PCR profiles can distinguish between types of organisms between them, but also show a set of fragments' characteristic for the individual genera. At the same time, the presence of a different number of identical-sized fragments at representatives of different species and genera of microorganisms talk about their closer or more distant genetic relationship.

In a visual readout of the resulting matrix, DNA profile (fingerprint) is difficult to consider all derived DNA fragments of a strain. Following the introduction of computer analysis that was possible. The purpose of our study was to construct a dendrogram using the method UPGMA. Resulting in REP-analysis dendrogram, according to the arrangement of the tested strains on the electrophoretic field grouping of the strains into eight main groups (Figure 9.3) is shown. Some of the groups include representatives of the same

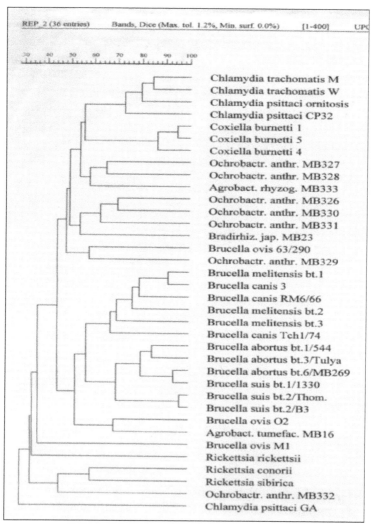

Figure 9.3 Dendrogram of strains Coxiella burnetii, Rickettsia (α-1 sub-division of Proteobacteria), α-2 sub-division of Proteobacteria, Chlamydia, and Brucella; REP-PCR; UPGMA.

genus and species, and othersinclude different types of the same or different genera, according to the previous classification of species.

Made on the basis of the dendrogram, analysis indicates that the similarity used in the study between chlamydial strains was 85% for Chlamydia. trachomatis (lines 1–2), and between them and C. psittaci ornithosis and C. psittaci CP32 (lanes 3–4), respectively, 72 and 79%. An exception was observed in C. psittaci GA (line 36), which has 30% similarity with the rest of chlamydia and phylogenetic stay far away from them.

In the group of Coxiella burnetii (lines 5–7), the reference human strain Coxiella burnetii 1 (Henzerling), phylogenetically stands much closer to Coxiella burnetii 5 terrain strain isolated from aborted cow (95% similarity) than to human isolate Coxiella burnetii 4 (86% similarity). Toward the other groups of strains, Coxiella burnetii show a genetic distance of 2 to 65%. Coxiella burnetii 1 and 5 are genetically the least distant from Br. suis bt.2 (2%) (lines 28–29), of Br. abortus bt.6/MB269 and Br. suis bt.1/1330 (3%) (lanes 25 and 26) and by Brucella melitensis bt.1 and Br. canis 3 (lines 17 and 18). For Coxiella burnetii, 4 genetic distance is smallest with Br. abortus, bt.1 and 3 (lines 23–24) and C. trachomatis (lines 1–2) –2%.

Rickettsiae characterized by its polymorphism show less genetic similarity. It was 36% in the reference human strain of Rickettsia ricketsii (line 32) causing Rocky mountain spotted fever and 58% between the Rickettsia sibirica (line 34), the causative agent of the North Asian spotted typhus—strain "Netsvetaev" isolated from the blood of a sick person and Rickettsia conori (line 33), responsible for Mediterranean spotted (Marseillaise) fever, and strain "Kardzhali-1," isolated from the tick R. sanguineus. Therefore, the last two strains phylogenetically sit on a branch.

It was found that brucella are very closely related to a DNA level and individual types are not always strictly separated. Brucella melitensis bt.1 and Br. canis 3 (lines 17 and 18) are phylogenetically on a branch as the similarity between them is 88% and at Br. abortus bt.6/MB269 and Br. suis bt.1/1330 (lines 25 and 26), it was 92%. Brucella suis bt.2/Thom. and Br. suis bt.2/B3 exhibit 97% similarity (lines 28–29). By their DNA profile, some of them come close to bacteria of α-2 sub-division of Proteobacteria. For example, Br. ovis 63/290 and Ochrobactrum anthropi MB329 (lines 15 and 16) exhibit 56% similarity, and Brucella ovis O2 and Agrobacterium tumefaciens MB16 (lines 29 and 30) exhibit 64%.

The data presented demonstrate that with the help of computer analysis, pinpointing the locations, structure, and interrelationships between organisms becomes possible, which will inevitably contribute to a possible change in the classification of microorganisms.

9.3 Computer Analysis of the RAPD Fingerprints (DNA Dactyloscopy)

In RAPD analysis, the primers used included OPL 04 and P5. The resulting DNA profiles separately with both primers were similar but not identical and contained between 8 and 23 fragments. Visually, the results of RAPD analysis with primer P5 were sharper, so we used them to compile the dendrogram (Figure 9.4).

Dendrogram structured according to the program PHYLIP, version 3.5s.

RAPD analysis revealed similarity of the studied strains Chlamydia trachomatis in 93% (lines 1 and 2), of C. psittaci CP32 and C. psittaci ornithosis

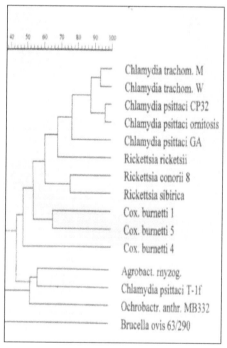

Figure 9.4 Dendrogram of strains of Coxiella, Rickettsia, Chlamydia, Brucella ovis, Agrobacterium rhizogenes, and Ochrobactrum anthropi. RAPD-PCR. Computer program PHYLIP, version 3.5s.

in 96% (lanes 3–4), and the similarity between the two groups was 88%. C. psittaci GA and C. psittaci T-1f (lanes 5 and 13) are phylogenetically more distant from the other types of Chlamydia as the similarity of them is, respectively, 77 and 56%. C. psittaci T-1f (line 13) exhibits greater similarity with Agrobacterim rhizogenes, which are located on the same branch of the phylogenetic tree (line 12).

When comparing these results with the dendrogram based on the REP analysis (Figure 9.3), the presence of the less similarity is established—85% in the first two strains, and 79% and 72% for the latter two. In REP-dendrogram, C. psittaci CP32 and C. psittaci ornithosis are of different genetic distance, while in the dendrogram based on the RAPD, it is of equal. The similarity of Chlamydia. psittaci GA with other Chlamydia is also much lower—as low as 30% at 77% in the RAPD analysis. Consequently, RAPD analysis is more suitable for the construction of the dendrogram to determine the genetic distance between strains of the genus Chlamydia—lines 1–5, 13 (Figure 9.4).

The same is observed when comparing the dendrogram of the studied strains Rickettsia. In RAPD analysis, the percentage similarity between Rickettsia conorii and Rickettsia sibirica, which are phylogenetically of equidistance (lanes 7 and 8), is larger (75%) than that of PCR analysis (58%). The similarity of Rickettsia rickettsii with the other strains was 68% in RAPD and 36% in the REP analysis (lane 6).

The greater sensitivity of RAPD, due to the higher resolution of the method in terms of Chlamydia and Rickettsia, makes it suitable to determine the genetic relatedness between the representatives of these two genera.

In the RAPD analysis of strains Coxiella burnetii is seen (Figure 9.4) that Coxiella burnetii 1 and Coxiella burnetii 5 (lanes 9 and 10) in the dendrogram are also of the same genetic distance, but the similarity between them is lower (64%) than in the REP analysis (95%). The same applies for Coxiella burnetii 4 (line 11), which exhibits 47% similarity with the remaining strains C. burnetii in RAPD and 86% in the REP-analysis. Accordingly, REP-analysis is more suitable for the construction of the dendrogram to determine the genetic similarity between the strains of the genus Coxiella, than RAPD analysis.

Used as control strains, Ochrobactrum anthropi MB329 and Brucella ovis 63/290 in REP-analysis are on the same branch of the dendrogram with 56% similarity, while in the RAPD analysis, they have a different distance and exhibit similarity, respectively, 52 and 28.

We find rather important results of our work in the field of PCR for detection and identification of Coxiella burnetii in strains of different origins.

This work certainly expands our diagnostic arsenal and our research spectrum in the field of Q fever [612, 613].

The qualities, opportunities, and prospects of PCR techniques in studies of C. burnetii and Q fever are discussed in several publications [311, 336, 451, 486, 488, 489, 493, 522, 554, 605, 614].

Very good results obtained by us under normal PCR with the primers CB-1 and CB-2 confirm the data of Stein and Raoult [489] for their specificity and effectiveness in detection and identification of strains of C. burnetii. Later, Kounchev [614] has also successfully used the specified primer pair in a modified conventional PCR. Our results in the identification of R. conorii, R. rickettsia and R. sibirica with primers Rr 190.70p and Rr 190.701n coincide with those obtained by Fournier et al [197] with the same primers. For identification of R. rickettsii, Eremeeva et al [182] included in the PCR primers Rr 190.70p and Rr 190.602n, amplified fragment of 532 bp and homologous sequences in the other two types of rickettsiae. We will emphasize once again that the adapted of our method for conventional PCR technique is sensitive and reliable for the detection and differentiation of the pathogen to the species level, but is not suitable for differentiation within the species between the representatives of one genus. For this purpose, we applied the above-described PCR techniques—REP-PCR and RAPD-PCR. Used as controls in the study members of the group ticks, spotted fevers, Rickettsia conori, Rickettsia sibirica and Rickettsia rickettsii, also for the first time in Bulgaria were investigated using PCR. As regards the other control groups of microorganisms in the country has prior studies on PCR in different versions—conventional, REP, ERIC, RAPD—on Chlamydia organisms from Martinov et al [335] and on brucellae and other bacteria—Tcherneva et al [506, 507]. The results of our experiments show that some of the above techniques of PCR, successfully used to differentiate, typing, and epidemiological studies of certain bacterial and chlamydial agents [535, 551, 566, 508, 335], are fully applicable and sensitive to Coxiella burnetii and rickettsias from the group of tick-borne spotted fevers. The results obtained by REP-PCR amplification fragmented DNA profiles and constructed dendrogram provided valuable information on the comparative analysis of C. burnetii and control groups. They not only allow differentiation of microbial species studied between them, but also show a set of fragments characteristic of different genera, as well as information for close or more distant genetic relatedness between them. In studies with RAPD was found that this type of assay is less suitable for phylogenetic studies of strains of the genus Coxiella, compared with REP-PCR. The method RAPD-PCR showed greater sensitivity based on higher resolution against chlamydial agents and rickettsiae, making it suitable for determining correlations and genetic similarity between representatives of these two genera.

In general, the different variants and modifications of the PCR techniques, characterized by a high specificity and sensitivity, are established as part of the diagnostic arsenal for proving C. burnetii and Q fever in animals and humans. In no case, however, they do not eliminate the long-known and widely applied other methods, but rather complement and confirm the results. In certain cases, polymerase chain reactions are the only credible alternative for diagnosis and scientific research. So, undoubtedly are expanding the possibilities of veterinary and human medicine for timely diagnosis, tracking, effective treatment, and control of Q fever as a zoonosis with a global distribution and importance.

10

Clinical Picture

10.1 Clinical Course of Q Fever in Animals

The clinical characteristics of Q fever in animals are an interesting and still insufficiently studied aspect. Overall, the undeniable existence of latent asymptomatic form of flow, disclosed by serological investigations, has unjustified perception of its dominant and almost only clinical condition of infected animals. Considering the Q fever as a unified for animals and humans widespread zoonotic disease, Migala 2004 [355] states that there is no full understanding of its diverse clinical manifestations. Undoubtedly, differences in symptoms in domestic animals deserve special attention [118]. In number of publications has been reported clinical disease in domestic animals, often with mass character [16, 45, 275, 336, 456, 573, 581]. Sanford et al. [457] describe "a storm" of abortions in 5 goat herds in Canada. Besides abortions, the authors registered stillbirths or newborn unviable kids dying in the coming hours or 1–2 days. Zeman et al. [581] reported mass miscarriages in sheep in Q fever in central and northern California with a range of 10–60% in the individual herds. For Q fever in different countries clinically manifested by pathology of pregnancy (abortions, premature births, dead, or non-viable offspring) in sheep, goats, and cows were presented data from Masala et al. 2004 [336], Italy; Hatchette et al. 2001 [228], Canada; Van der Lugt et al. 2002 [531], The Netherlands; Aldomi et al. 1998 [71], Jordan; Khaschabi et al. 1996 [275], Austria; Yoshie et al. [573], Japan and others. In the earlier publications, Bulgarian authors described miscarriages in sheep and goats as the main manifestation of Q fever [43, 44, 47]. According to Woldehiwet 2004 [564], the main clinical manifestation in domestic ruminants is the late abortion. Often abortions occur suddenly without prior symptoms [457]. In other cases, there was a slight-to-moderate discomfort, decreased appetite, and decreased milk secretion [573]. In some infected herds of cows are established a number of cases of births of calves with low weight [244]. Reproductive disorders and infertility in cows with Q fever are other forms of clinical manifestation [499, 445, 561, 380]. The mammary gland of domestic ruminants is a common location for the reproduction of C. burnetii and the development of acute, subacute, and chronic inflammation with release of the pathogen into the milk [118, 126, 254, 554]. According to Byrne [118], excretion of the agent with the milk secretion of lactating cows may acquire chronic in nature and can last for months or years. It can definitely say that we need more in-depth research on mastitis caused by C. burnetii in animals as a separate clinical form of Q fever.

In connection with the the question of the release of C. burnetii from the sick animals, a number of authors have found that the placentas, fetuses, and uterine-vaginal secretions are heavily infected and contaminate the environment with large amounts of the infectious agent. In the acute phase of infection, regardless of the degree its clinical expression, the presence of C. burnetii can be detected in the blood, lungs, spleen, and liver [343]. The infection in the animals under natural conditions often becomes chronic, which is accompanied by prolonged excretion of the causative organism in the feces and urine [126, 343, 410].

There are reports of Q fever in lambs clinically manifested with mass respiratory diseases [45, 249] and pneumonia in adult goats in herds where abortions are simultaneously detected [43]. So far, research on pneumonias and other respiratory and flu-like conditions in domestic animals caused by C. burnetii is limited and represents a promising direction in the infectious pathology of the animals.

In flocks of sheep, Q fever can occur with severe clinical syndrome—sudden onset of fever (42°C), numbness and difficulty in movement, mucous-purulent conjunctivitis and rhinitis, and reduced milk secretion to a full stop after 3–4 days. The temperature values were reduced slightly to 41°C for 3 days and remain within these limits until the 8th day. Many animals die. Other sheep weaken strongly and recover hard [16]. Mölle et al. [367] describe mass disease in sheep in Germany with a clinical picture including ataxia, anorexia, and progressive exhaustion. The animals had normochromic aplastic anemia. Postmortem C. burnetii is detected directly by IF and LM and by the isolation of the agent from the central nervous system (CNS) and parenchymal internal organs.

According to Literak and Rehaček [304], the development of clinical disease Q fever in cattle depends on the virulence of the respective strain. Established in the Czech Republic and Slovakia latent form of infection in the cows and the complete absence of clinical diseases is due to poor circulation of virulent or even avirulent strains. The authors found that the bovine strain C. burnetii isolated from them is not virulent for sheep when trying to infect experimentally [304].

Clinical signs in other domestic mammals and birds affected by spontaneous Q fever are insufficiently explored. Single evidence suggests that in spontaneously infected dogs can be observed births of unviable puppies those died in a short time [115]. According to the OIE, 2004 [380], except in ruminants, miscarriages, stillbirth, retained placenta, endometritis, infertility, and births of feeble offspring with low weight can occur in cats, dogs, rabbits, and other species. The normal births in infected cats in some cases are not accompanied by abnormal clinical condition; in other cases are observed mild-to-moderate general signs preceding the birth—depression alternated with anxiety, appetite disorders, and hyperthermia. The direct contact with birthing cats often leads to infections in human beings [289, 399, 349]. C. burnetii can be isolated from the genital secretions of cats with vaginitis and endometritis [373]. Miscarriages in cats caused by C. burnetii infection were confirmed etiologically by Daoust et al. [153] and Higgins and Marrie [242]. Respiratory disease Q fever in cats has not been studied. Experimentally infected cats develop fever and become lethargic [380].

Experimentally induced clinical conditions in animal models to study the pathogenesis and pathology of acute and chronic Q fever—guinea pigs, white mice, rabbits, monkeys, etc.,—are described in the Chapter 11.

10.2 Clinical Observations in Sheep

10.2.1 Latent Form

Under natural conditions, latent course of Q fever is common. For disclosure of latent form are used serological methods. The absence of visible clinical signs in the presence of antibodies against C. burnetii in a herd put the requirement for careful analysis of the primary serological data and systematic clinical and epidemiological surveillance in the coming months and years.

We found that in cases where the percentage of serologically positive sheep is small and at the same time titers were mostly lower, the apparent circulation of less virulent strain was a prerequisite for lasting balance between the agent and the body of the animal, respectively, for long-lasting latent inapparent course.

In the absence of clinical diseases in a herd, where the percentage of serologically positive sheep is medium-high or high, with predominantly moderate and high titers, asymptomatic period of Q fever was shorter and gradually developed visible disease manifestations. This transformation from latent form to the clinical symptomatology is more recent, and the more virulent strain C. burnetii is circulating in the herd. An important indicator in this respect is the behavior of the animals during the lambing season, the exact registration of the abortions, and other health failures in this period and laboratory confirmation of the etiology of the occurred pathologic conditions.

In some cases, especially in flocks of sheep which have not been the subject of prophylactic serological screening, latent infection with C. burnetii was disclosed following clinical observations and etiological proof of the prevalence of Q fever in humans from the same farm or village.

The latent form in some cases is found completely independent in the herd or farm. In other ovine herds, some of the infected sheep show any signs and the others are latent infected but asymptomatic. These two categories coexist with clinically healthy seronegative animals at the time. Thus, in the same flock, we found various serological and clinical status of the infection with C. burnetii. Upon continuation of targeted clinical epidemiological and etiological studies on Q fever in these herds had revealed the dynamic nature of the epizootic process and its clinical manifestations, including oligo-symptomatic and asymptomatic latent forms [597].

10.2.2 Abortions, Premature Births, Stillbirths, and Not Viable Lambs

Under natural conditions observed number of cases of abortions—single, in a few animals or mass outbreaks. As a rule they appeared in the second half of the pregnancy, but most often in the last 1–2 weeks. Sometimes abortions are not preceded by signs of disease. In other cases, depression establishing two or three days before abortion increases the body temperature to 41°C, anorexia. After abortion, these disorders may disappear after one to three days. In a part of miscarried sheep watched retained placenta with subsequent development of endometritis or severe metritis, followed by an infection that causes the death of most of them.

Besides abortions in sheep of active foci of Q fever, we observed premature births, which had a different outcome according to the general condition of the newborn lamb and the effectiveness of measures taken to survival [597].

Often, the pathology of pregnancy in infected sheep manifested with births in normal physiological period of offspring of dead or sickly lambs underweight who died in the next 48 hours. During such births registered a fever (up 41.5°C), tachycardia, shortness of breath, rhinitis, conjunctivitis, lack of appetite, and reduced milk secretion.

We conducted a comprehensive study of the experimental infection of three pregnant ewes inoculated intravenously with infectious suspension of Bulgarian strain "Chilnov" with a titer10^6 ID_{50}/ml, cultivated in YS of CE [597]. Infection performed in different periods of pregnancy: ewe D1 at 99th day; ewe D2–71th day; ewe D3–45th day. Animals showed clinical signs of disease on the 5th–7th day after inoculation—internal body temperature increased to 40.9°C as the fever continued for 2–3 days. On 12th to 13th day, we registered a new temperature rise by 0.6–0.7°C followed by normalization of the 14th to 15th day. Besides biphasic temperature reaction, the sheep showed depression, thirst, and decreased appetite, especially expressed in the febrile period. There were still conjunctivitis, increased tear secretion, and accelerated breathing. The outcome of developing infection with C. burnetii is shown in Figure 8.7: premature births of not viable lambs on the124 day of pregnancy, i.e., 25 days after inoculation (ewe D1) or on the131 day—60 days after the infecting, in ewe D2. Infecting with C. burnetii of the third ewe D3 resulted in a pathological outcome of the 110th day of gestation, respectively, i.e., 65 days after inoculation—premature birth of a dead offspring and a second non-viable lamb. Premature live lambs were with cachexia, arthritis, and ataxia and up to 24 hours ended in a fatal outcome.

The obtained results show that the experimental inoculation with a virulent strain of C. burnetii having ovine origin of the sheep in the first pregnancy (1 month and 15 days, and 3 months and 7 days) resulted in an infection with severe clinical consequences—premature termination of pregnancy and dying of the fetus. Serological response, pathomorphological changes, and electronic microscopic findings are presented in the relevant sections (see for more details).

10.2.3 Pneumonias in Sheep

Under natural conditions, pneumonia caused by C. burnetii is observed in adult sheep and lambs of different ages [601].

Etiologically proven cases of Q fever in sheep with clinically manifested pulmonary inflammation found in 20 farms from different regions of the country. The scope of the disease varies from sporadic cases or small groups of animals to mass involvement of entire herds. In epizootic of Q fever in November/December 2004 in the region of Shumen, Bulgaria, along with the established abortions, 80% of the sheep in five neighboring flocks of standard size developed pneumonic form of the disease.

The general clinical characteristics of the observed pneumonias in older sheep includes frequent persistent cough, typical of inflammation of the lungs, finding in percussion and auscultation, wheezing, increased vesicular breathing, bronchial breathing, and rapid and difficult breathing movements. At the beginning of clinical disease, animals are febrile, but for some of them remained elevated to temperature at a total clinical picture of pulmonary inflammation, which in some cases was a bronchopneumonia and in others, interstitial pneumonia type. In systematic follow-up of the hyperthermia in individual sheep establishes fever with persistently elevated and slightly fluctuating temperature values. Most often, however, it was found febrile illness with peaks and troughs of the temperature curve over a wide range.

Overall, we registered three groups of clinical course in adult sheep:

a. *mild-to-moderate* pneumonia with relatively preserved general condition;
b. *medium-heavy* form with an average of strength deterioration of the general condition;
c. *severe* with strong deterioration of the general condition, phenomena of total intoxication, recumbency, respiratory, and heart failure. Some of these animals died, despite having taken remedial measures.

In our observations, also we found adult sheep with chronic respiratory inflammations—bronchopneumonia or atypical interstitial pneumonias and chronic bronchitis with periodic acute exacerbations.

In weaned lambs aged 12–14 months have seen massive pneumonias in a holding, while in another—both in weaned lambs and younger lambs, located next door. Clinical onset manifested by sluggishness and slowness of movement increased body temperature, reluctantly eating to a complete denial of food. The pulmonary inflammation was accompanied by deaf cough, shortness, and difficulty in breathing. During auscultation can be heard wet wheezing. The moderate forms of pneumonia ended in recovery. Upon moderate and severe pneumonias, recovery was slow. In some animals, the clinical course took adversely development and ending with death. From pneumonic lungs of such lambs isolated the C. burnetii.

In lambs aged 1–6 months in five farms from different areas with etiologically confirmed Q fever looked at acute bronchitis, pneumonia, and bronchopneumonia with varying severity of clinical course. We observed disease usually in a moderate to moderately severe form that ended with a

slow recovery. Some of the animals were sick with severe continuously deteriorating general condition, respiratory and cardiac insufficiency, intoxication,
and death.

In another section of this monograph, we presented data on an experimental infection in sheep and two lambs after nasal or tracheal inoculation
with an isolate C. burnetii from the lungs of lambs suffering from pneumonia
(Chapter 8, Section "Serology in pneumonias in sheep"). Along with tracking
the dynamics of the convincing serological response (Figure 8.12) undergo
clinical observations [601]. The three animals—one adult sheep, which was
not pregnant, and two lambs aged two months—developed Q fever with
clinical symptoms of pneumonia.

In lambs, the disease had a sharp development, especially highlighted
in tracheal infection. Rise in body temperature in tracheal inoculated lamb
(A2) began on the 24th hour, and in lamb A1 inoculated nasally at 48 hour
(Figure 10.1). This is followed by a gradual increase in temperature to

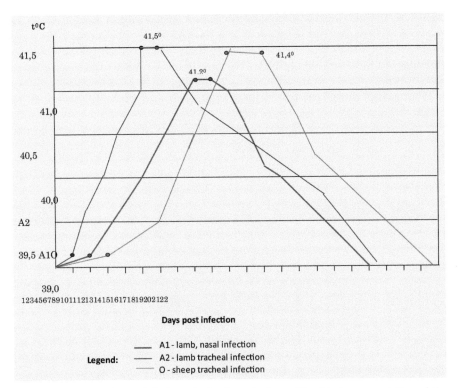

Figure 10.1 Experimental Q fever in lambs and sheep. Graphical representation of the
temperature reaction after nasal and tracheal infection.

41.5°C /fifth day/ in A2, respectively, to 41.2°C in A1 /eighth day/. After holding the maximum values for one day, started their downward until complete normalization after 9–10 days (Figure 10.1). The general condition of the animals was deteriorated both in febrile phase and after it: depression, rigors, anorexia, recumbency, pain on palpation of the chest, wheezing, cough deaf, hard vesicular breathing, and signs of partial thickening in the lungs.

The clinical picture of sheep infected tracheally was similar, but with more moderate, subacute course, though the expressive and more protracted temperature reaction, which ended with the normalization of the day 22 (Figure 10.1).

10.2.4 Flu-Like Form in Sheep

This form is characterized by the appearance of low or moderate fever pronounced as the only clinical syndrome: increased internal body temperature, depression, partial or complete loss of appetite, and stiff movements. Febrile illness lasts from 6 to 12 days. In individual sheep, also we found serous-catarrhal inflammation of the mucous membranes of the eyes and nose, accompanied by intensified conjunctival and nasal secretions. The prognosis is quite favorable in the absence of complications caused by the infection itself with C. burnetii or due to secondary intervention of other etiologic factors of contagious or non-contagious nature [601].

Flu-like form of Q fever in the case of poorly manifested fever and insufficient experience of the livestock personnel and veterinary staff incorrectly can be attributed to the latent form of the disease.

10.2.5 Mastitis in Sheep

Observed by us outbreaks of mass mastitis in sheep caused by C. burnetii is suddenly appeared by rapidly developing picture of high fever: temperature rise to 41.5–42°C lasting 48 hours, loss of appetite, and increasing depression. In the mammary glands of lactating animals began inflammatory processes occurring with significant tempering, redness, tenderness, and edematous. These phenomena affect posture, gait, and movement of sheep that were cautious, stiff, and sparing. Milky discharge quickly changed its normal color, quantity, and odor, becoming the first in serous-catarrhal liquid with yellow-gray color, and later in dirty brown foul-smelling exudate. From the clinical onset of the severe inflammation, the amount of secretion continuously decreases to almost complete stop. The most commonly observed bilaterally affected udder. Taking quick remedial measures in the initial phase of the disease leads in many cases to prevention heaviest complications and slows healing of the mammary gland but usually the secretion of milk does not resume until the end of this season of lactation. When antibiotic therapy is delayed or it is ineffective the inflammatory process deepens and the color of the mammary gland becomes dark red, blue or violet. The infection acquires generalized character and some of the animals die.

After the acute stage, in the worst-affected sheep the secretory function of the mammary gland did not recovered or remained incomplete. This physiological inferiority and the related economic inexpediency led to the removal of the animals from breeding and production [602, 616].

10.3 Clinical Observations in Goats

10.3.1 Latent Form

In a spontaneous Q fever in goats, the latent course of the illness is detected by serological test on goats of different age groups [597, 594]. The prophylactic or targeted serological screening of clinically healthy animals in areas, farms and herds of unknown epidemiological status, or in known farm foci allows us detecting the infection and its categorization as follows:

a. latent form of Q fever in herds without clinical manifestations from farms where the disease was first recognized;
b. a latent form of the disease in herds without clinical diseases that are in immediate or in close proximity to other infected herds;
c. latent Q fever of animals in herds with etiologically proven clinical forms of the disease.

The range of latent asymptomatic infected goats varied within wide limits. Thorough clinical supervision, combined with periodic serological test, allows us to track the dynamics of the spread of infection with C. burnetii in the continuous absence of visible disease events and for registration of appeared later overt clinical manifestations. In this regard too are indicative the observed changes in clinical status of some latently infected goats in a season of births, which in the next gestation have abortions due to infection with C. burnetii. So, carefully and systematically performed clinical etiological surveillance of Q fever was useful in prognostic respect [597].

10.3.2 Abortion in Goats, Premature Births, Stillborn, and Non-viable Offsprings

Clinical and pathological conditions related to gestation are typical manifestations of Q fever in goats [597]. Abortions are usually in the last two months before the time of birth and often appear in the last 2–7 days. In our studies, we observed no reference clinical behavior of the animals prior to the abortion. In some cases, the goats showed no previous signs and abortions occur suddenly in single goats and small groups. The number of animals affected in some herds grew rapidly. In other cases, abortions were preceded by feverish condition with varying severity of symptoms—elevated body temperature, rigors, depression, changes in appetite, and physical activity. Separate goats showed signs of conjunctival inflammation—hyperemia, lacrimation, and rarely bright fear. The fever continued from 2 to 5 days, followed by abortion. Part of the abortions registered in the midst of the feverish condition. When disposing of the entire placenta with aborted fetuses, usually the goat recovered immediately or within two or three days. Retention of the placenta, which often observed, was leading to inflammation of the genitals, sometimes with serious complications due to septic phenomena and general intoxication. The features described above were also concerned about stillbirths occurred prematurely or in a normal period of pregnancy.

Premature births of kids alive in the last month of pregnancy, especially very close to its closing date, had a different outcome, according to the general condition of the newborn, respectively, extent of possible injuries due to infection with C. burnetii.

In some goats affected by Q fever had births in normal time but newborn were underdeveloped and not viable that led to their death within 1–2 days. If signs preceding these births, these were similar to the above or register only short-term temperature rise. General ill health postpartum observed more frequently in the cases of retained placenta.

Found out that the clinical symptoms in goats with abortions and other similar expressions of pathology of pregnancy manifested with the particularity that the deterioration of the general condition in its three degrees (mild to moderate, averaging more pronounced and severe) occurs not only as a preliminary, but also as subsequent abortion or birth phenomenon.

This feature also observed in infected animals which have normal delivery. Obviously not only as pathological gestation with its consequences (abortion, etc.), but also the normal delivery in infected goats results in activation, exacerbation, and generalization of the infection with C. burnetii [597].

10.3.3 Respiratory Form in Goats

Inflammation of the lower airways, acute bronchitis and pneumonia, is the another feature of the course of Q fever in goats [597]. Often these conditions are sole or dominant clinical syndromes in the development of the disease. Affecting both young and adult animals of both sexes usually occurr as disease outbreaks and rarely as sporadic cases. Sometimes in the same infected goats, we observed both respiratory disease and abortion.

The clinical onset is manifested by fever, which lasts 2–4 days and in individual animals, up to 6 days. It is accompanied by hyperthermia (40.7–41.7°C), depression, reduction or disappearance of appetite, and significant fatigue, leading to bedsores to some goats—appearing serous and later serous-catarrhal and mucopurulent nasal discharge.

The inflammatory process in some cases reaches the bronchi and develops acute bronchitis without further penetration into depth. In other cases, infection with C. burnetii affects the lung parenchyma and the interstitium, which clinically manifests with bronchopneumonia or pneumonia, most often from interstitial type.

Upon total clinical picture of pneumonia, the general events worsen. Frequently were established tachycardia, shortness of breath, chest tenderness on palpation, deaf and deep cough. Auscultatory most often bilateral were heard wheezing, crackles, increased vesicular breathing, and bronchial breath. In the absence of etiological diagnosis and lack of treatment with appropriate effective antibiotics, the pneumonia in goats caused by C. burnetii often leads to death.

10.3.4 Flu-Like Form in Goats

In herds of goats, established etiologically Q fever has also observed a slight ongoing influenza-like illness affecting all age groups [597].

In newborn kids, mostly between the 6th and 14th day after birth, we observed febrile illness, which lasted about a week. The internal body temperature of newborn is increased significantly. The conjunctival and the nasal mucosa get red, edematous, and excrete abundant serous discharge. The mucosa of the oral cavity also becomes hemorrhagic and swollen a little. Signs of inflammation of the trachea, bronchi, and lungs missing are seen. With few exceptions, the general condition of the kids was not affected. The appetite remained preserved and the physical activity remained unchanged.

In adult goats, the influenza-like form of Q fever was manifested with transitional mild and often moderate fever, hyperthermia, depression, drowsiness, muscle pain, and reduce but not complete loss of appetite. Milk secretion in part of the sick goats was reduced. Conjunctival mucosa was red and swollen, mostly bilateral and produce serous–mucus secretion—likewise the observed acute rhinitis with profuse nasal secretions. The outcome of the disease was favorable and usually occurs a spontaneous recovery after 7–8 days.

10.3.5 Mastitis in Goats

In active foci Q fever in goats, relatively often are detected clinically manifested mastitis. Particularly susceptible to infection with C. burnetii are the young goats at first delivery and put into the herd newly purchased animals which more often become ill of Q fever and the clinical course of them is predominantly acute.

The development of mastitis is preceded by a week febrile stage (40.3–41.6°S), accompanied by changes in behavior and the overall status of goats—malaise, lethargy, decreased appetite, and lagging behind the herd grazing. Mammary gland in most animals was visibly inflamed, but generally, unlike sheep, severity of the disease is milder, without severe and long-lasting organic and functional damages. Observed are mild-to-moderate hyperemia and edema of the udder and tenderness to the touch. Milk secretion decreased rapidly or gradually to full suspension. Himself milk secretion in many of the sick goats showed no visual standouts changes, especially in milder cases. Some affected animals however had visibly changed milk with gray and white or whitish yellow color. From such secretions without difficulty was isolated C. burnetii [594]. With the passage of the febrile phase, along with an overall improvement in the condition of the animals we observed reduction and resolution of the clinical signs of mastitis. Mammary glands regain its color and turgor, the secretion of milk in amount and organoleptic properties recovered over the same lactating. In all cases of mastitis caused by C. burnetii, the question of the duration of excretion of the pathogen in the milk in clinical and convalescent stage requires special attention and control measures [597].

10.4 Clinical Observations in Cattle

10.4.1 Latent Form

In our extensive research on virtually all over the country, we found that the latent infection with C. burnetii, asymptomatic and mildly course of Q fever in cattle are common, and their establishment is possible by using serological reactions. In some settlements and cattle farms, the discovery of serologically positive animals without clinical signs of disease was the first signal for the circulation of the agent and the associated need to start measures to control Q fever as a disease of economic and zoonotic importance. In subsequent clinical observations and sero-epidemiological studies considers the scope of the latent-infected cattle, their age, sex, and production purposes, as well as the likely impact of other factors such as potential activators of the latent infection and its transformation into clinical disease. In that respect, emphasis was concerned to pregnancy, birth, and postpartum of the individual cows. [597, 598].

First detection of latent Q fever in cattle in some cases achieved after purposeful serological tests based on initially received clinical, epidemiological, and etiological data on diseases of Q fever in people who raise cows in the same area.

Often, latent infection revealed in herds without clinical disease found in immediate or close proximity to farms where Q fever is symptomatic.

In the active foci of Q fever with many positive sero-reagents and etiologically proven clinical cases in early, total and convalescent phase, both latently infected animals were found. Some of the latter were constantly inapparent, and other animals had suffered from clinical disease, but remain asymptomatic infection carriers.

10.4.2 Abortions in Cows, Premature Births, Stillborn, and Non-viable Calves

Abortion and related clinical conditions in cows were found in 45 settlements in different areas of the country. Usually, individual animals or small groups were affected, and the tendency to increase or decrease the number of abortions in a holding was depending on a number of factors, primarily—the activity of the infection, the virulence of the strain, the timely etiological diagnosis, and effectiveness of the performed therapeutic and general measures.

Abortions were observed in the second half of pregnancy, most often in the last 1–2 months. Often cows aborted 1–2 weeks ahead of schedule. In many cases, no signs do not augur sudden loss of the fetus. Other animals showed mild or moderate discomfort, vaginal mucous discharge, medium-high hyperthermia, reluctantly meals, or reduction of the pick feed. As an exception observed prior pronounced signs of anxiety, distress, fever, and anorexia. After abortion under detached placenta, for several days the cows had strong genital discharge, but the general condition rapidly improved. The retention of the placenta, especially for long periods, leads to rapid rotting process, heavy odor, inflammation of genitals, and progressively deteriorating general condition of the animals.

Premature births per alive or dead calves also ran in the presence or absence of pre-existing symptoms similar to those described for the abortions. Newborn calves of these cases, especially when almost complete wear unaffected or slightly harmed general condition, survived with adequate care. In heavily damaged vital functions, born prematurely calves died in the next 7 days.

Also registered deliveries within a time limit a fully worn, but dead calves and sickly, underweight calves with poor viability. Such animals came to death within 1–2 days. In these clinical–pathological cases, the preceding symptoms were similar, but less pronounced compared to abortion. Retained placenta in cattle were also a risk factor for the occurrence of complications similar to those described above.

In some cows after abortion or pathological birth at incontinence placenta, we concluded the development of general illness with moderately or severely impaired general condition (fever, temperature rise, anorexia, fatigue, etc.). Most likely, this is due to the aggravation and generalization the infection with C. burnetii as a result of the pathological uterine activity of these animals [597].

10.4.3 Clinical Signs of the Dry Cows after Experimental Infection

Two of the dry cows infected intravenously with C. burnetii and traced their serologic response and clinical status for 160 days (Chapter 8).

For the effectiveness of the infection testifies the dynamics of CF-antibodies against the agent of Q fever depicted graphically in Figure 8.20 (See Chapter 8 "Serology"). The graph shows also that the two infected animals react with a three-day increase in internal body temperature—between the fourth and sixth day in a cow No. 1 and between the sixth and eighth day in cow No. 2. Besides this clinical signs, we observed moderate signs of depression alternating with anxiety and reduce appetite and physical activity. These symptoms resemble to some extent flu-like form of Q fever, described in small ruminants.

In cow No. 1, registered late recurrence (120 days) of febrile reaction accompanied with mild-to-moderate worsening of the overall condition of the animal. Re-febrile phase preceded the new, more moderate increase in serologic titer, followed by the closing date of the experiment—its 160 days [597].

10.4.4 Pneumonias in Calves

Affected were calves aged 2–4 months. The incubation period was about one week. Clinical disease beginning with serous-catarrhal rhinitis displayed by nasal discharge. The body temperature of the animals was slightly higher (40–41°C) and the other is in the range 41–42°C. Appears sonorous cough that with the development of the disease quickens. The appetite decreases. Further development of the disease was acute or subacute and was characterized by more quickly or gradually unfolding picture of lung inflammation by simultaneously affecting the bronchi (bronchopneumonia), or more often with atypical course (interstitial pneumonitis). Depending on the type of the pneumonic process, the findings in percussion and auscultation were different. In bronchopneumonia the cough was sharper, with higher tone and greater force. Clearly could hear wet wheezing and crepitus. In interstitial pneumonia, the cough was muted and wheezes, dry or barely, discovered. Often in auscultation could be heard bronchial breathing. Also we observed signs of breathlessness. With continued deterioration of the calves grew stronger dyspnea and cardiac function deteriorated due to toxic myocardiopathy and insufficiency. Such animals died or because of the unfavorable prognosis slaughtered by necessity. The clinical development for subacute cases was favorable and illness signs gradually disappear [597].

10.4.5 Endometritis, Metritis, and Infertility

Studies show that C. burnetii infection in cows with reproductive disorders occupies a certain share in the etiopathogenesis of these conditions [591, 615–619].

The clinical manifestation of the concept of "reproductive disorders" is not homogeneous, but rather a collection of several disease syndrome representing themselves distinct forms of clinical manifestation of infection with C. burnetii. The most common general consequence of these diseases is sterility in cows.

Acute and chronic inflammations of the uterine lining (endometritis) usually appeared as a result of retained placenta (Retentio secundinarum) of cows infected with C. burnetii. This abnormality observed in both abortions and births of dead or non-viable calves and under normal births with carriage and excretion of C. burnetii with the fetal membranes and uterine-vaginal discharge at the puerperal period.

Endometritis in their acute phase showed a uterine-cervical secretion, originally from the serous type, passing later in serous-mucous and mucopurulent.

The genital discharge has a different character—permanently or intermittently and more abundant or scarce. In the secretions obtained by uterine-cervical swabs proved the presence of C. burnetii [597, 615].

Colposcopic observations revealed inflammation of the vaginal portion of the cervix and endocervicitis, often of granular and follicular type, and ectopic changes—vascularization and vesicular lesions of the cervix and easy bleeding when handling. The development of endometritis is obviously a result of ascent infection from the cervical canal of the cow. In chronic inflammation, the visible clinical signs are the same, but more mildly affected. The repeated attempts at artificial insemination or mating naturally with bulls were unsuccessful. This state is named by the breeders "frequent repeats."

In some cases, after a successful antibiotic therapy is achieved cure of inflammatory processes. The result is overcoming the temporary inability to fertilization—in other case reached to permanent infertility, leading to severe economic losses.

Endometrial inflammation, mainly acute, can be complicated by development of metritis representing severe inflammatory disorder of the uterus. Besides the observed external signs such as abundant mucopurulent or purulent discharge with a bad odor, often occurring complications of a general nature—systemic infection with phenomena of intoxication and unfavorable prognosis. A typical example of this is observed by us, septic metritis in a cow from large active agricultural focus of Q fever (Piperevo, Kn, Bulgaria). After a miscarriage on five months of gestation caused by infection with C. burnetii and retention of the placenta, followed an acute inflammation of the endometrium, which gradually covered all tissues of the uterus. With the development of metritis there were general signs such as hyperthermia, denial of food, severe fatigue, and inability to standing in an upright position. The septic phenomena and the general intoxication led to severe disturbances in heart and throughout the body. A fatal outcome occurred on the day following the onset of the common morbid phenomena [597].

In herds with reproductive problems witnessed other cases of metrites in cows as a result of infection with C. burnetii that despite the severity of inflammation in the uterus during vigorous and prolonged therapeutic interventions were being managed and achieved clinical and microbiological recovery. Some of these animals were restoring their reproductive abilities. Quite a few from the cows suffered, however, were infertile, which required their removal from the herd.

Reproductive disorders were also found in sero-positive for Q fever cows without clinical evidence of endometritis and metritis, but having disturbances in sexual cycle (estrus) and decreased milk secretion. In the protracted and deepening of these conditions, part of the animals concerned passed in the category of infertility.

10.4.6 Mastitis in Cows

Clinical observations of mastitis performed at 72 sero-positive for C. burnetii cows originating from four farms where investigated total of 235 animals with diseases of the mammary gland. Thus, the estimated range of Q fever in cows with similar pathology was 30.6% [597]. Performed at 72 sero-positive for C. burnetii cows originating from four farms were investigated total of 235 animals with diseases of the mammary gland. Thus, the estimated range of Q fever in cows with similar pathology was 30.6%. [597].

In 37 cases (51.4%), it is moderately expressed inflammation of the udder of serous and catarrhal type. The mammary glands of these cows were mild-to-moderate hyperemic, tempered, and painful to the touch. Milk secretion showed a decrease in quantity. In the milk, changes were observed due to its admixture with serous and catarrhal exudate.

The remaining 35 cows (48.6%) were affected by severe mastitis with catarrhal-purulent or purulent character manifested in a pronounced hyperemia, cyanosis, high tempering and painful udder causing changes in posture and gait. Milk secretion sharply decreases and stops. Milk secretion is strongly altered and has become virtually inflammatory exudate.

As a complication of severe form of the acute mastitis in some cows observed generalization of the infection with symptoms of high fever, fatigue, depression, anorexia, and phenomena of total intoxication.

The outcome of the disease was depending on the severity of the course, the period of implementation, and the effectiveness of the antibiotic treatment. Lighter and moderate forms lend themselves more easily to therapy, and in many cases, there was a complete recovery of the mammary gland and its production. Certain severe forms of mastitis by vigorous treatment with antibiotics also responded, but udder slowly made whole, and the secretion of milk coming back in the next lactating.

Long untreated and very serious mastitis had dubious prognosis regarding the full restoration of the mammary gland and often these animals are culled from dairy farming. In generalized form of infection with C. burnetii, the prognosis is unfavorable.

10.4.7 Syndrome of Prolonged Reduction of Milk Production

The reduction of milk secretion as impaired physiological process and economically significant consequence often accompanies the above-described clinical forms of Q fever in cattle.

In certain cases of etiologically proven infection with C. burnetii do not detect other obvious clinical and pathological manifestations besides lowering the amount of milk produced. The decline in milk production was in varying degrees and for extended periods of time. This condition is designated as *Syndrome of prolonged reduction of milk production* [620].

Independent examination of this issue separately from the latent form of Q fever is necessitated by the fact that at last there are no clinical signs including reduced milk secretion.

10.4.8 Clinical Signs in Some Laboratory Animals

Rabbits. The clinical picture traced in five groups of domestic rabbits experimentally infected with C. burnetii isolated or directly demonstrated in different clinical forms of Q fever in sheep and cattle [597].

The experimental animals pretreated orally with tetracycline in 3 days, in order to inhibit the possible latent infection. The inoculations with infectious material made after one week in order to eliminate residual concentrations of the antibiotic. We performed experimental infection of rabbits that were serologically negative for antibodies to C. burnetii and Chlamydia.

The implementation of infection with C. burnetii in rabbits demonstrated through immunofluorescence and light microscopic pathogen detection in the spleen, liver, uterus, and placenta and also based on the positive serological test results for the presence of specific antibodies in blood sera.

First group. Ten male rabbits were infected intraperitoneally with 10 000 ID_{50}/ml of laboratory strain "Chilnov" isolated from the placenta of aborted sheep and passaged 125 times in YS of CE. The infectious titer of the strain was 10^{-6} ID_{50}/ml by Reed and Muench.

On the second day after inoculation, the internal body temperature started to rise and reached 40°C on the seventh day and then normal. During this period, the experimental animals were bristling and had sluggishness in movements, conjunctival redness, slight acceleration, and more frequent respiratory movements. The dynamics of specific antibodies generated was between 1: 256 and 1: 1024.

Second group. Four pregnant rabbits /10–22 days / were infected intraperitoneally twice every three days with 3 ml of blood (first injection) and

6 ml (second injection) from ewes in stage of bacteraemia after experimental infection with the above ovine strain of C. burnetii "Chilnov".

The clinical picture in these animals is characterized by the start of the temperature rise in third–fourth day and a maximum of 41.2°C on the seventh day, depression, bristled, slow movements, slight conjunctivitis, and moderate tachypnea. All four rabbits abort between the sixth and ninth day after infection. Complement fixation titers against C. burnetii ranged from 1: 64–1: 256.

Third group. Three male rabbits were infected intravenously with 3 ml of strain "Tutrakan" of C. burnetii, isolated in chicken embryos from milk of sheep suffering from mastitis. The animals reacted with signs of fever. The beginning of the temperature rise was recorded 24 hours after inoculation, and its maximum is on the fourth day. Complement fixing specific antibodies appeared on the seventh day (1: 8–1: 16) and progressively increased to levels 1: 256–1: 512 between the third and fourth week.

Fourth group. This group consisted of three males and three pregnant rabbits infected with concentrated and purified suspension of cotyledons and placental tissue of aborted cow. The inoculation material contains C. burnetii in a high multiplicity(10^8).

After intravenous injection, all animals developed fever lasting 5–6 days. Pregnant rabbits aborted between fifth and eighth day after inoculation. Serological response and this group were expressive.

Five group. Three rabbits injected intra-testicular at a dose 2 ml of yolk suspension strain of C. burnetii, isolated from the genital discharge of cow with septic metritis.

Clinically, after an incubation period of three days, we concluded hyperthermia (41.4–41.6°C) lasting 6–10 days and inflammatory edema of the scrotum and surrounding tissue (orchitis and peri orchitis). In the blood serum of the rabbits accumulated specific antibodies in significant titers.

Guinea pigs. Clinical observation of Q fever in guinea pigs held after experimental infection with C. burnetii—directly from clinical and pathological material (placenta, fetal parenchymal organs, and milk secretions) or isolates of the agent most often in CE, from sheep, goats, and cattle with different clinical forms of Q fever [597]

In many cases, we used the method of intraperitoneal infection. Inoculating of 10% suspensions of placental tissue, cotyledons, and internal parenchymal organs from fetuses in abortions in sheep and births of dead or unviable offsprings led to clinical illness with the symptoms of fever. The temperature climbed after incubation of two to seven days and ranged from 39.80 to 41°C.

The duration of temperature increase was from four to eight days. In some guinea pigs, hyperthermia was the only sign of disease. Other animals showed anxiety, followed by depression, numbness, and pricking. Mortality is in the range of 30 to 60%.

The subcutaneous inoculation with the same starting materials from sheep leads to infection of guinea pigs, but after a longer incubation period and with moderate signs. Some of them responded with a local inflammatory reaction of the skin and subcutaneous connective tissue at the injection site.

After intraperitoneal inoculations with the agent of ovine abortion and related conditions, propagated in chicken embryos or spleen of guinea pigs was developed a generalized infection similar to the described above clinical picture. In all cases, the intensity of the clinical manifestations was dependent on the concentration of the agent in the infectious inoculum.

We followed also the clinical manifestations of guinea pigs infected intraperitoneally with C. burnetii, isolated in CE after inoculations of material from the lungs of sheep with pneumonia. The inoculations were performed with native yolk suspensions. The incubation period from an average of seven days was followed by a febrile phase (40–41°C) and systemic disease with a mortality of about 40%.

The experiments with infectious material from sheep with mastitis conducted by intraperitoneal inoculations directly with milky exudates and separately with concentrated and purified coxiellas from an isolate in CE obtained from the same infectious milk secretions. In both cases, we observed marked pathogenic to laboratory animals, clinically manifested by high fever, changes in general condition, and death rate of 50%.

The series of experiments for intraperitoneal infections in guinea pigs also included the strains PM (mastitis in goats), HP (abortion in cows), MP (isolation from placenta of normal birth cow), and Henzerling (reference of human origin). The clinical characteristics of induced Q fever included four indicators:

a. an average incubation period (days): 16.2 (PM); 6.1 (HP); 13.1 (MP), 8.3 (Henzerling);
b. medium-term hyperthermia (days): 3.2 (PM); 4.6 (HP); 3.5 (MP); 6.3 (Henzerling);
c. the average value of the increased internal body temperature (t°C): 39.6 (PM); 40.1 (HP); 39.7 (MP); 40.5 (Henzerling);
d. mortality (%): 0 (PM) 83.3 (HP); 50 (MP); 25 (Henzerling).

Diseases of the guinea pigs were accompanied by accumulation of CF-antibodies against C. burnetii as the highest titers registered with HP.

The above four strains of C. burnetii of caprine, bovine, and human origin were tested by inoculation in the parenchyma of the testes. Induced disorders clinically were manifested by temperature-scrotal phenomenon—hyperthermia and orchitis. The temperature reaction continued from 7 to 14 days. The inflammatory edema of the testicles and the tissues of the scrotal bag led to massive enlargement and tenderness of the scrotum, which hampered the movements of animals. The lethality in this clinical form of infection with C. burnetii was significant.

White mice. Clinical signs in white mice observed after their experimental infection with C. burnetii isolated from the disease in sheep, goats, and cattle [597].

Clinical sign after intracerebral infection. Young white mice weighing 7–8 grams inoculated under ether anesthesia intracerebrally with 0.1 ml of 10% three types of suspensions: from aborted ovine placentas and cotyledons and from fetal parenchymal organs and infected yolk sac of chicken embryos. In advance with the help of electron microscopy demonstrated the presence of C. burnetii in starting materials—inoculating suspensions with a high concentration of the agent, 10^6–10^8.

The introduction of the pathogen in the brain of mice resulted in an evolution of the disease causing massive mortality after four to eight days. The specific nature of lethality was confirmed by light and electron microscopy of preparations from the brain. We observed the presence of large intra-cytoplasmic inclusions of C. burnetii (see Chapter 4).

Intracerebral inoculation of mice with suspensions of YS of CE infected with C. burnetii isolated from milk exudate was obtained from sheep with mastitis-induced disease with a strong deterioration of general condition and significant mortality between the third and ninth day. The etiological confirmation of infection was similar as described above.

Clinical signs after intraperitoneal infection. By this method, inoculated are several groups of white mice with infectious material (YS of CE) of abortions and related conditions in sheep, mastitis in sheep and goats, and normal births in the same animals.

The first group of mice divided into two subgroups as one injected i.p. with a suspension of aborted fetuses and placentas having a rich accumulation of C. burnetii. Animals from the second subgroup were inoculated with a suspension of an isolate in CE, also with a high multiplicity of the agent. We found that both types of suspensions in 1 ml dose do not induce well-developed clinical signs, and approximately half of the mice were asymptomatic. Lethal cases were very rare. Part of the mice remained completely resistant to infection. For realized infection with C. burnetii at clinically poorly manifested cases and especially in the clinically inapparent animals, we judged by the enlargement of the spleen and the pathogen detection in microscopic preparations, as well as by the detection of specific antibodies in the serological tests.

Clinical signs follow intranasal infection. The nasal applications of C. burnetii, isolated and cultivated in chicken embryos of sheep with mastitis, performed with two types of suspensions: native not concentrated and concentrated and purified.

Clinical signs following intranasal infection. The nasal applications of C. burnetii, isolated and cultivated in chicken embryos of sheep with mastitis, performed with two types of suspensions: native not concentrated and concentrated and purified. The latter contains very rich accumulation of the infectious agent—10^8 –10^9. Both inocula introduced at nasal route, caused infection of mice with different clinical course: latent inapparent or barely perceptible features in the use of native suspensions and shown visibly general and respiratory signs when using the concentrated suspensions. Breathing becomes more frequent and difficult (pneumonia)—appetite decreased. In some mice, we observed a tremor of the body.

The nasal inoculations of white mice with yolk suspensions of isolates of C. burnetii in mastitis in goats and abortions in cows induced prolonged respiratory diseases in some of the laboratory animals. Other mice showed less developed respiratory and general symptoms tending to spontaneous resolution. The rest of the mice were completely asymptomatic. The presence or absence of infection caused by C. burnetii demonstrated after autopsy and microscopic studies of preparations of lungs and spleens.

10.4.9 Analysis of the Clinical Data in Animals

The question of the clinical manifestations of Q fever in animals has always been to the attention of clinicians and microbiologists, but even today there is no complete understanding of the diverse clinical manifestations of the disease [355]. The presence of multiple hosts of C. burnetii assumes no uniform response of the individual species, including with respect to the clinical symptoms. Byrne [118] points out that the differences in the symptoms exhibited by infected domestic animals deserve special attention and research efforts to identify and characterize.

Data from the literature stands, generally speaking, *three groups of clinical forms of Q fever in animals*: (a) latent inapparent; (b) pathology of pregnancy (miscarriages, still births, and non-viable offsprings); (c) pooled third and expanding category, including a number established but insufficiently studied conditions—puerperal acute and chronic inflammation of the reproductive system and related reproductive disorders and infertility; mastitis; respiratory diseases; severe generalized clinical syndromes often fatal; and syndromes characterized by prolonged reduction of the milk production.

It should be noted that in the attempts to characterize the Q fever in some publications is exposed unsubstantiated concept of latent infection as predominantly and almost exclusively clinical condition of the animals concerned.

In our studies based on large clinical and experimental material, observations were focused mainly on three types of domestic ruminants (cattle, sheep, and goats), recognized as a major source of infection for humans. Clinical observations conducted in all other species—the subject of etiological studies on Q fever. Also registered are clinical signs in laboratory animals serving as biological models for studying the behavior of the agent and the infection caused by it.

In sheep, goats, and cattle, the latent infection with C. burnetii, poor clinical course, and asymptomatic form were common and were revealed by serological reactions. The latent form is neither uniform nor static. In some cases, these animals are found in herds with no apparent clinical disease. In other instances, inapparent latently infected animals coexisted with clinically ill infected with C. burnetii ruminants.

In the active foci of Q fever with multiple affected animals at different stages of infection and diverse clinical manifestations, we found *two types of latently infected individuals* (a) permanently asymptomatic at the time of infection; (b) suffered from clinical disease, but remained inapparent infection carriers. Our system observations showed that the clinical status of some latently infected animals is transformed into visible symptoms. In particular indicative in this respect are the periods of pregnancy and delivery and the puerperium, in which often drew activation of the latent infection leading to clinical manifestations—abortions, unviable newborns, detentions of placenta, inflammation of genitals, hard, and sustained release of C. burnetii in the environment [46, 96, 97, 105, 381, 386].

Summarizing the results of the sero-epidemiological and clinical studies conducted on a large territory indicates that the latent infection with C. burnetii occurs most often in pigs, horses, and buffaloes—species with limited epizootiological and epidemiological importance at this stage. Among the main agricultural reservoir and source of infection for the animals and peoplecattle, sheep, and goats—the latent form of Q fever is set mostly in the bovine population, followed by sheep and goats. In these animals, however, the designation "asymptomatic" is largely of notional nature, as is often accompanied by changes in important physiological indicator—the secretion of the mammary gland expressed in its continuous decrease, resulting in significant economic damage.

From the foregoing it appears that to the latent form of Q fever should be approached carefully taking into account the facts of its dynamic nature and potential for conversion into clinical disease and the continued excretion of the agent in the environment. In all cases of detection of asymptomatic sero-positive for C. burnetii animals in a herd must avoid unilateralism; carefully conducted history questionnaires, clinical and epidemiological analysis, including historical data and targeted clinicalobservations, as a rule reveal clinical diseases that often they were transported at the expense of other etiologic factors of infectious and non-infectiousnature, in the absence of planned or aimed laboratory investigations for Q fever.

Our clinical observations of etiologically confirmed cases of naturally acquired Q fever revealed many cases of miscarriages, premature births, stillbirths, or non-viable offspring in cows, sheep, and goats in many locations. These findings were complemented and enriched by the results obtained in the complex study (serological test, electron microscopy, clinical, and postmortem) on an experimental Q fever in pregnant ewe-lambs. The clinical picture is rich. It is characterized by a particular characteristic biphasic reacting temperature, symptoms of acute respiratory condition, and premature births of dead or non-viable lambs. The registered experimental results are also useful for the differential diagnosis of Q fever and chlamydial abortion in sheep. In ewe-lambs affected by infection with chlamydia observed mass abortions without previous or with minor clinical signs, while ewe-lambs affected by Q fever show pronounced signs of fever and have abortions relatively rare but often give birth to unviable lambs. Received by us data in spontaneous or experimentally induced Q fever confirmed the pathogenic potential of C. burnetii to cause abortions and related conditions with great economic and epidemiological significance, which was also reported in other publications [43, 44, 47, 228, 244, 275, 336, 457, 531, 564, 581].

Too telling are our observations on respiratory diseases in sheep, goats, and cattle with naturally acquired Q fever. In small ruminants, we observed severe pneumonias affecting all ages of both sexes with varying severity of clinical course and disease ranging from mass involvement of whole herds to groups of animals or sporadic cases. Also observed are bronchitis and bronchopneumonia in lambs, kids, and goats, and chronic recurrent bronchitis and pneumonia in adult sheep. The clinical and etiological findings in the natural infection with C. burnetii were convincingly confirmed by experimental reproduction of respiratory disease in lambs and sheep after nasal and tracheal infection with the causative agent. Information about the etiological role of C. burnetii in mass respiratory diseases in lambs was presented by Hore [249] in Australia and Ognyanov et al. [45] in Bulgaria. In another publication [43] is described Q fever in flocks of adult goats, manifested clinically with pneumonias and abortions. In cattle, the most vulnerable age group were calves of 2 to 4 months. These animals were affected by acute and subacute bronchopneumonia and interstitial pneumonias with characteristic clinical picture. Undoubtedly, the disclosure of lung infections in calves, as a result of infection with C. burnetii, expands the clinical spectrum of Q fever in cattle.

Besides pneumonias and bronchitis, we observed a flu-like form of Q fever in sheep and goats, which describes for the first time in the literature [601]. The recognition of this clinical condition in herds with proven etiological Q fever was without difficulty if there is a significant or moderate pronounced signs of fever and catarrhal occurrences from the ocular and nasal mucous membranes. In poor manifested fever and minimal or almost missing ocular and nasal secretions, influenza-like form of Q fever may erroneously be attributed to latent form, especially with insufficient observation and experience of live-stock and veterinary personnel. In cattle were not conducted targeted clinical observations for possible disclosure of flu-like shape in spontaneous Q fever. Very interesting results received, however, in experimental infection in the dry cows after i.v. infection with the agent. Realized infection with C. burnetii, as evidenced by the dynamics of the specific antibodies, was also manifested with clinical disease resembling to some extent the above-described flu-like form of Q fever.

The excretion of milk has long been known fact that focuses attention from epidemiological point of view [118, 351, 385, 554]. There is evidence that the excretion of the agent in the milk of lactating cows may acquire chronic nature and to continue for months and years [118]. In contrast, the very clinical event of inflammation of the mammary gland remains in the shadow of the epidemiological implications of the spread of the pathogen by this mechanism, all the more that its excretion can be carried out in the absence of clinical evidence of mastitis. In other cases, such as goats [43], mastitis is regarded as part of the overall clinical syndrome of Q fever, including deterioration in the general condition of the animals, fever, conjunctivitis, rhinitis, decreased or absent appetite, and late abortion considered a main symptom.

We focused our attention on clinically manifested mastitis in sheep, goats, and cows running as a separate, self-occurring disease affecting large numbers of animals. Particularly acute disease runs in sheep where delay in etiological diagnosis and treatment with antibiotics leads to generalization of the infection with C. burnetii and some animals died. Suffered severely affected sheep remained dysfunctional due to incomplete or permanently discontinued milk secretion. Such outbreaks of mastitis in sheep are described for the first time. In cows and goats also be established severe forms of mastitis, and more moderate or mild forms. In all cases, mastitis etiological associated with C. burnetii led to severe economic losses due to reduced milk production and removal of severely affected animals of breeding and production because of the economic appropriateness of keeping them.

Economic importance also has other clinical forms observed by us and laboratory proven as a result of the Q fever—retained placenta, endometritis, metritis, reproductive disorders, and infertility found in the cows, sheep, and goats. Note that the data submitted for Q fever associated with genital inflammation, reproductive disorders, and infertility studied in more detail in cows are the first of its kind in Bulgaria [615, 616, 597]. The significance of the problem at the international level is emphasized in publications reflecting his research in Germany, Denmark, Switzerland, France, Italy, Hungary, Cyprus, Canada, USA, India, and Japan [288, 332, 380, 445, 499, 520, 561, 617–619].

For a more complete study of the infection with C. burnetii in dogs for which were received convincing serological data are necessary clinical observations, particularly in shepherd dogs as well as farmed dogs or raised at homes where this task could be perform easier. The same applies to cats. So far these observations are limited. Although not numerous literature reports describe clinical events in Q fever in dogs and cats as manifested by abortions, stillbirths and newborn unviable, retained placenta, endometritis, and infertility [115, 373, 380].

From the foregoing it appears that Q fever in domestic animals has a rich and varied clinical characteristics, part of which is the latent asymptomatic form having dynamic nature and potential to activate and transform into clinical disease. Ignorance of the clinical particulars of the disease, coupled with the failure to take preventive or focused serological tests, is the basis of incomplete and insufficient diagnosis of Q fever in certain areas and farms.

Permanent elements in our studies were also the clinical observations of laboratory animals (guinea pigs, white mice, and rabbits) used for experimental infections in the study of different aspects of the infection caused by C. burnetii. These observations were an important indicator of the effectiveness of the chosen route of infection and clinical criteria for the characterization as well as for the pathogenic effect of the agent from different origin.

They complement and extend the results obtained in isolation and cultivation of the agent, immunological activity, pathogenesis, pathological morphology and ultrastructure of C. burnetii and the affected tissues and cells of the host. All these confirms once again that used laboratory animals are valuable biological models for studies on Q fever in animals and humans [2, 43, 82, 199, 286, 322, 343, 476]. Experimentally induced abortions in domestic rabbits and overall clinical response in these animals are grounds to conclude that C. burnetii infection lies a potential risk of introduction and spread among rabbits kept in private yards and industrially farmed as an additional source of meat for the population.

10.4.10 Clinical Course of Q Fever in Humans

In humans, the clinical course of C. burnetii infection is manifested in three main varieties: asymptomatic form; acute Q fever; and chronic Q fever. In the three main groups of clinical forms are marked different variations of the expression: subclinical, atypical, affecting various organs and systems [26, 72, 140, 192, 209, 255, 426, 599, 609]. The incubation period ranges from 2 to 40 days (average of 2–3 weeks) depending on the concentration and dose of inoculum C. burnetii [129, 118, 343]. The shortening of the incubation period is directly proportional to the increasing number of the pathogen cells trapped in the body. According to the CDC, USA, 2005 [126], about 50% of the persons infected with C. burnetii show signs of clinical disease.

Asymptomatic infection (asymptomatic seroconversion) is revealed serologically. In some cases, individuals without signs representing smaller groups of people have inhaled small amounts C. burnetii of contaminated clothing or materials similar to those described in Canada [321]. In other broad sero-epidemiological studies in identified epidemics of Q fever unless clinically manifested cases are revealed simultaneously multiple asymptomatic serologically positive individuals or ones with poorly developed signs [173]. Like animals and in humans is described persistent C. burnetii infection in asymptomatic condition [188, 201, 325]. From placentas of asymptomatic women infected by 1 to 6 months have been achieved isolations of *C. burnetii* in phase I [325, 201].

Acute Q fever. Most cases begin with sudden high fever, severe headache with frontal or retrobulbar location, and chills. Also described are myalgia, angina, sweats, non-productive cough, chest and abdominal pain, vomiting, diarrhea, depression, fatigue, anorexia, weight loss ($\geq$7 kg), arthralgia, stiffness in the neck, and neurological signs [52, 126, 228, 253, 424, 478].

These signs are often nonspecific and resemble those seen in severe influenza, so crucial is the laboratory diagnosis [126, 509].

From 30 to 50% of patients with symptomatic Q fever develop *atypical pneumonia*. Another common clinical manifestation is *hepatitis*. Often the infection occurs only as a febrile illness accompanied by very strong headache.

The incidence of the most frequent clinical forms is not the same in different geographical zones of the world and across the different countries. For example, in France, in a study of 1383 cases of Q fever has been found that acute forms were 1070/77.3% and the chronic – 313/22.7% (Raoult et al., [426]). Acute diseases are occurring in 40% as hepatitis; 20% as pneumonias and hepatitis; 17% as pneumonias; 17% as febrile conditions; and 1% myocarditis, pericarditis, and meningoencephalitis. Chronic forms of 259 (82.7%) were endocarditis and rarely chronic hepatitis, pericarditis, osteoarticular lesions, and others [426]. In Australia, the dominant form of acute Q fever is a *"febrile disease"* [485, 375]. The pneumonia is the dominant form in Canada [326] and so on. The severity of the course of pneumonias etiologically linked to C. burnetii is different—mild, moderate, and severe. The latter is without or with complications (respiratory failure or discontinuation of respiratory function, encephalitis, severe kidney or heart damage, myocarditis [330]. The mortality rate is below 3% [343]. Hepatitis is of granulomatous type [320, 478]. In the majority of cases is established hepatomegaly with pain in the right upper quadrant [378]. Liver enzyme levels are increased— AST (aspartate aminotransferase) and ALT (alanine transferase)—from 2 to 3 times above normal in 50 to 75% of patients, and similar increases in the level of alkaline phosphatase are scored at 10–15% of cases. There are reports of fatalities due massive destruction of the liver tissue, leading to hepatic coma [95, 165].

Other clinical forms of acute Q fever in humans are pericarditis, myocarditis, brain inflammation, and skin rashes. For pericarditis in Q fever reported a number of authors [176, 205, 316, 364–366, 439, 529]. Particularly often the disease was established in Spain, but it is possible omission of such cases in other countries due to lack of targeted search. The same goes for myocarditis regarded as a rare but life-threatening form of acute Q fever [132, 474, 494]. The latter is disclosed by electrocardiogram, or on the basis of the clinical observations in cases expressed, accompanied by tachycardia, hypoxemia, and acute heart failure, which can lead to death [494]. There have been reports of individual cases of encephalitis, meningoencephalitis, and encephalomyelitis appearing in the later stages of acute Q fever [482, 190]. They are described as well as skin rashes—pink spots or purple-red papules [172, 517].

Marrie [323] states and other more *unusual clinical manifestations* observed in acute Q fever: mediastinal lymphadenopathy, pancreatitis, thyroiditis, optic neuritis, hemolytic anemia, erythema nodosum, extrapyramidal neurological disease, epididymitis, and orchitis. The overall mortality in acute forms of Q fever is 1–2% [126].

Chronic Q fever is represented primarily by *endocarditis* [138, 189, 209, 426, 477, 492]. This serious disease occupies 60–70% of all chronic diseases etiologically related to C. burnetii. The most commonly affected aortic heart valves, to a lesser degree—the mitral and rarely—tricuspid valve [298, 406]. From the tissue of the valves is isolated C. burnetii. The majority of the affected persons have had a previous heart valve disease. Endocarditis and the other forms of chronic Q fever may develop in patients with acute infection in periods of 1–20 years after infection with C. burnetii [126]. Risky patients for chronic Q fever are immuno-suppressed patients with transplants of organs, cancer patients and those with chronic renal failure [126, 421]. Although more rarely, endocarditis caused by C. burnetii was registered in AIDS patients [112, 422]. The findings of the cited French authors were confirmed recently in the USA.

The clinical manifestations leading to the diagnosis of endocarditis caused by H. burnetii are diverse and nonspecific. This provides an explanation for the significant delay in diagnosis for months and even years, which is essential for the prognosis of the disease [343, 316]. Goffin et al., [209] from the European Bank for cryopreserved heart valves in Brussels reported that the reason for the rejection of transplanted valves in 5.4% of cases over a 10-year period of observation is persistent infection with C. burnetii.

Another clinical form of chronic Q fever in humans is the *vascular infections* [112, 188, 179, 196, 424]. In samples from the tissue of blood vessels, particularly aneurysms of the aorta, through the isolation and PCR has been demonstrated C. burnetii [196]. The diagnosis is made by serological detection of the characteristic of chronic Q fever levels in the first-phase antibodies against C. burnetii. Clinically in most patients, there is a severe inflammatory syndrome, fever, abdominal pain, weight loss, very fast reaction of erythrocyte sedimentation, high serum levels of fibrinogen, and C-reactive protein [196, 343]. These signs are nonspecific and it is therefore necessary to carry out serological test for Q fever.

Chronic infection with C. burnetii can also occur with *boneand jointsymptoms*: osteoarthritis [416, 392], osteomyelitis [179, 147], infection of the aortic pseudoaneurysm with vertebral osteomyelitis [400, 179]. The infection is established in immuno-suppressed adults. More often are affected children having established contact with infected goats and cats [147]. Clinically, In these cases have been observed Inflammation of the hip joint and spondylitis. In the blood serum of patients accumulate specific antibodies from the first phase, with high titers. From the affected tissues is achieved isolation of the causative agent [550].

Rare clinical forms of chronic Q fever are the chronic lung infections—pneumonic fibrosis [69], inflammatory pseudotumor of the lung [262, 299], and chronic hepatitis without simultaneous finding of endocarditis [572].

The mortality rate in patients with chronic Q fever is high—23 to 65% [126, 400].

Also described is a clinical syndrome characterized by persistent fatigue and exhaustion in patients who are in convalescent period after acute Q fever [83, 84, 227, 317]. In the cases observed in Austria and Great Britain are marked tiredness to exhaustion sometimes lasting months and years, joint and muscle aches, sweats, enlarged and tender lymph nodes. Like domestic animals, C. burnetii plays a etiological role in the pathology of pregnancy in humans—abortions [166, 423, 428, 494]; premature births [94, 494, 304]; and intrauterine death [201]. Those conditions are often the result of inflammation of the placenta from which the agent is isolated. C. burnetii infection is proven also in fetuses [494]. The infected pregnant women have fever and develop influenza such as illness [304] or atypical pneumonia [297, 494]. Some of those affected were asymptomatic. Often there is a pronounced thrombocytopenia [494].

11

Pathogenesis and Pathological Morphology

Issues related to the pathogenesis of Q fever in animals and humans are partially defined. The development and the manifestations of the disease are dependent on many factors [124]. Pathogenetic factors for the host include the type, immune status, presence, and absence of competing infections. Probably, the role of the host in the pathogenesis of acute Q fever is based on the action of other unknown factors. On the side of the host the factors influencing the evolution of Q fever to chronic form are pregnancy in mammals and women [85, 494]; immunosuppression in laboratory animals and humans [124, 421, 422]; prior diseases of the heart valves in guinea pigs, mice and humans [82, 287, 419].

Significant factors for the pathogenesis of Q fever are the infectious agent—strain or serotype; virulence and pathogenicity for the species concerned; routes of infection (respiratory, alimentary, intraperitoneal, etc.); and dose of the inoculum [124, 159, 286, 322, 579]. Campbell [124] indicates that the pathogenesis and pathology of Q fever can be understood to some extent based on the tissue affinity and behavior of C. burnetii in cells. The pathogen survives and proliferates in lysosomal vacuoles and is incorporated in phage lysosomes [87]. Targeting C. burnetii to specific types of cells (lymphoid or myeloid) suggests that there is the presence of specific receptors. In a study, Honstettre et al. [246] found that lipopolysaccharide of C. burnetii is involved in phagocytosis of the organism and in the inflammatory response by receptor 4 that resembles barrier. Despite research efforts in general, genetic and cytological aspects of the pathogenesis of Q fever have not been studied satisfactorily.

Upon previous scant information about the pathogenesis of the disease in naturally infected animals was formed the following hypothetical scheme. Coxiella burnetii hits the body through the respiratory tract and digestive tract and in the blood-sucking ticks. After penetration into the blood, bacteremia develops, resulting in a temperature rise and formation of antibodies. In the internal parenchymal organ, due to septic and toxic phenomena in the splenic follicles develop hyperplasia, and in the liver, cardiac muscle, mammary gland, and CNS degenerative changes appear. Reticuloendothelial system is activated. Tropism of the agent to the placenta leads to inflammation of the placenta and abortion and to the respiratory tract—bronchitis and pneumonia. In the acute phase, C. burnetii is found in the blood, lungs, spleen, and liver. The infection often becomes chronic, which is accompanied by sustained release of the pathogen in the feces and urine. Placenta, amniotic fluid, and fetuses are severely infected. In acute or chronic inflammation of the mammary gland is established the excretion of the agent with the milk [118, 244, 288]. In pathological investigations in abortions, premature births and stillborn offsprings in sheep, goats and cows were observed macroscopic and histopathological changes [294, 360, 456, 457, 581]. Placentas are thickened and pale with hemorrhagic or granular areas and exudates resembling cream and having yellow-brown or gray-brown color on the surface. Cotyledons are dark red and dotted with necrotic areas. There is a picture of purulent placentitis and massive damage to the cotyledons of inflammatory necrotic processes, while fetuses, according to some authors, do not show macroscopically visible changes [456] and others reported their livers are enlarged and sealed [361].

In a number of publications were described microscopic findings in abortion and related conditions in domestic ruminants: placental vasculitis, accompanied by infiltration of mononuclear cells, neutrophils or eosinophils, with intravascular thrombus in individual cases [360]; purulent placentitis and hypertrophy of the chorionic epithelium, the cytoplasm of which accumulates many coccus-like cells of C. burnetii [294]; strongly expressed necrotizing placentitis [581]; granulomatous hepatitis and non-suppurative interstitial pneumonia in fetuses or non-viable offsprings [361, 581].

Some types of laboratory animals (guinea pigs, white mice, rats, hamsters, rabbits, and monkeys) are used as animal biologic models for the study of pathogenesis and clinical pathological features of acute Q fever in humans and farm ruminants. Observations show that the severity of clinical signs and pathological changes is unequal and laboratory animals can be completely asymptomatic or suffer from symptoms of fever and granuloma formation. In other cases, they may be affected by serious illness leading to death, especially when using highly infectious inocula of C. burnetii—over 10^6 infectious units [343].

The guinea pigs after intraperitoneal or intranasal inoculation increased body temperature after 1–2 weeks ($\geq 40°C$) have the presence of the pathogen in the blood continued for 5–7 days and later form specific antibodies—a second phase after the 15th day; a first phase—in the second month after the infection. Several days after inoculation of the agent are detected pathological findings in various organs. Often, the spleen and mesenteric lymph nodes are enlarged. Macroscopically in spleen, liver, and bone marrow were observed granulomas and in lung tissues—infiltrates of mononuclear cells. During the convalescent period, specified abnormalities regress without residual findings. This type of dynamics of the infection with C. burnetii in guinea pigs is very similar to acute Q fever in humans [343]. Experimentally infected guinea pigs often become latently infected, but the induction of immunosuppression by treatment with adrenocorticosteroids [475]. X-ray irradiation of the whole body [476] or cyclophosphamide [82] leads to the reactivation of the disease. Such exacerbation of Q fever was observed in people with endocarditis after administration of corticosteroids [293]. La Scola et al. [286] examine the relationship between the mode of infection and pathological lesions in guinea pigs. Intraperitoneal inoculation of C. burnetii leads more often to granuloma formation in the liver than to intranasal introduction of inoculum. Conversely, the findings of cellular infiltrates in the lungs prevail in intranasal infection [286]. Used high doses of the pathogen ($\geq 10^5$UI) can cause myocarditis with characteristic pathological features [286].

Upon the white mice methods of inoculation of C. burnetii also influences the form of the pathological manifestations [322]. Splenomegaly and hepatomegaly were detected only after intraperitoneal infections achieved with strain Nine Mile and two other human strains (MPZ—placenta; Q229—endocarditis) and also with the strain isolated from the uterus of a cat. The microscopic observations revealed hepatic granulomas predominantly in mice infected i.p. and to a lesser extent in nasally inoculated animals. And in mice, findings of interstitial pneumonia prevail considerably after intranasal infection compared to those seen in i.p. inoculation [322]. Some of the inoculated mice develop inapparent latent infection. Like guinea pigs, treatment with steroids and X-ray radiation can lead to the exacerbation of the disease with clinical signs and postmortem changes [475, 476].

Less frequently used animal models for studying the acute Q fever are hamsters, rats, and monkeys. In hamsters, the main location of C. burnetii is spleen [500]. Gonder et al. [210] reported studies on pathogenesis and pathological morphology of Q fever in monkeys. Following aerosol infection, animals develop pneumonia between the 4th and the 7th day, and the presence of the agent in the blood is established between the 7th and 13th day. Histopathological findings are interstitial pneumonia and subacute hepatitis. Ethological studies indicate the presence of C. burnetii in the lungs, liver, spleen, heart, kidney, testes, and specific serum antibodies in the first and second phases [210]. Upon experimental infection of pregnant rabbits were induced stillbirths, but there are no postmortem studies of these conditions.

For studies of chronic Q fever in humans (endocarditis) have been developed four animal models: pregnant white mice [343]; immunocompromised mice with cyclophosphamide [82]; guinea pigs previously damaged by electro-coagulation heart valves [287]; and rabbits with intracardiac catheters [363]. Upon i.p. inoculation of pregnant mice with a strain Nine Mile is developing chronic infection, as manifested by births of mice with low-weight and histological lesions typical of endocarditis [343]. In immunocompromised mice challenged with high doses, C. burnetii establishes infection, disseminated in various organs, including the heart valves [82]. For the study of endocarditis with typical histological changes of the affected valve and the simultaneous isolation of the agent from the blood they have been used two other animal models [287, 363]. An important precondition for taking these tests is that used animals do not develop endocarditis in natural infection with C. burnetii.

In humans, in similarity with the presented records of the animals, including animal biologic models stand a few moments in the hypothetical concepts of the pathogenesis of Q fever. In the acute form of the disease, there are a variety of clinical manifestations—from asymptomatic latent infection to a lethal disease as a result of damage in pneumonia, hepatitis, encephalitis, myocarditis, or other more unusual locations and manifestations [170, 171, 286, 329, 428]. Probably, the clinical signs and pathological changes in the corresponding organs and systems are in direct or indirect dependence of the already mentioned and other factors: ways to deploy the agent in the human body; virulence of the causative organism; dose and immune status of the body; and other factors for the host [286, 293, 329, 330]. Breathing infected with C. burnetii aerosols leads to lung inflammation as a dominant clinical syndrome while alimentary route of infection with raw milk is connected mainly with the development of granulomatous hepatitis [194, 329, 355]. The dose of the infectious inoculum is essential for the severity of the clinical manifestations and pathomorphological lesions [286]. It allowed the opportunity for geographic variations in the clinical and pathological manifestations of Q fever due to genetic differences in the strains of the agent [343]. Byrne [118] chematizes pathogenetic events at Q fever in humans as follows. Coxiella burnetii infection usually results from the inhalation of infectious aerosols. The pathogen is phagocytic cells of the host, mainly not stimulated macrophages. This process is not dependent on energy, but is probably a result of the contact of the pathogen with the receptor.

Following phagocytosis, inside the phagolysosomal conditions are created stimulating the multiplication of C. burnetii, as at this stage is appearing the initial damage to the host cell. In parallel with the multiplication of the cells of C. burnetii, the cytoplasm is filled with them and injuries of cell organelles are growing progressively to lysis of the host cell. The infectious agent is disseminated in the body as a result of its free movement in blood plasma carried by circulating macrophages [118]. According to Mege et al. [354], the immunocompetent patients with acute Q fever develop transient bacteremia with C. burnetii as the pathogen is found intracellularly in the circulating monocytes, but not free of plasma. The authors express the suggestion that in patients having the defect of the heart valves and/or persisting presence of the pathogen in the blood due to immune deficiency, the infection with C. burnetii monocytes in the bloodstream can attach to damaged endothelial surface of the valve. This leads to the accumulation of C. burnetii cells and endocarditis [113, 354]. On the other hand, the amount of pathogen-infected monocytes detected in endocarditis associated with Q fever is too small, which probably means that in the pathogenesis of the disease are involved other factors, including the so-called immune complexes, regularly found in chronic Q fever [354]. The establishment of small quantities of infected monocytes in endocarditis is likely to be related to the microbicidal action of these cells based on the gamma interferon (IFN), limiting the intracellular multiplication of C. burnetii [158, 160, 161, 260] as well as with reduced transendothelial migration of monocytes infected with the pathogen [260].

According to Zusman et al. [585] between the systems involved in the pathogenesis of C. burnetii and Legionella pneumorphila, there is a similarity.

11.1 Pathomorphological Investigations in Domestic Ruminants with Etiologically Proven Q Fever

11.1.1 Macroscopic Picture

In pathological examinations of sheep pathology of pregnancy (miscarriage, stillbirth, and newborn non-viable lambs) based on Q fever acquired under natural conditions or after experimental infection, there were little visible macroscopic changes in the internal organs. The uterus is enlarged and swollen, and in the area around caruncles is diffuse redness.

The aborted fetuses are well developed, fresh, and slightly edematous. External signs of decay, maceration, and mummification are missing. The subcutaneous connective tissue is swollen. In the parenchymal internal organs were observed hyperemia and hemorrhage. In the abdominal cavity has an increased amount of fluid with a reddish color.

Upon dissection of stillborn and not viable lambs was observed that the liver and spleen are hemorrhagic, enlarged and have rounded edges. On the surface of the lung pleura were observed single petechial hemorrhages. The mesenteric and mediastinal lymph nodes are enlarged and juicy, with a wet-sectional surface. The meninges are hyperemic and have single hemorrhages on their incision surface [621].

The fetal placenta is hemorrhagic, edematous, and thickened with lost transparency. Cotyledons are swollen and hyperemic and in their surface were detected off-white taxed. The chorioallantois near the cotyledons is steeped in a gelatinous matter. In the same area stand strong blood-filled vessels. In the majority of cases, the chorion is diffusely thickened, folded, and has a dirty-red color. In some places, the surface is dotted with fibrinous–necrotic deposits with gray-white color [621].

Such macroscopic findings observed in goats and cows with abortions, stillbirths, and non-viable offspring [597].

In pathological investigations of pneumonia at dead sheep and lambs of different ages in the lungs observed inflamed areas with red-gray or gray-violet color in the form of single spots in the earlier stages of the disease, or diffuse large areas in advanced massive pneumonia. The lesions are unilateral or bilateral. Interlobular connective tissue is expanded. In individual animals are established bronchitis and peribronchitis. Mediastinal lymph nodes are enlarged (lymphadenitis). Inflammation of the pleura was found very rarely. In other organs usually are not found macroscopically visible lesions. One exception was the spleen, where it is often seen areas of inflammation with gray color [621].

In goats, macroscopic picture of respiratory diseases of the lower respiratory tract is similar. Compared to the sheep, more often can be found affected bronchial and peribronchial tissue. Regional lymph nodes are inflamed.

In calves observed similar pathological signs of focal pneumonia or lobar pneumonia type. In rare cases, it was observed simultaneously pericarditis.

In fatal cases of mastitis in sheep caused by C. burnetii in the mammary glands after cutting, we found strong hyperemia with separate cyanotic and hemorrhagic areas. The subcutaneous connective tissue and the interstitial connective tissue are edematous. The mucous membranes of milk ducts and milk tank are hyperemic and swollen. Sometimes are found pinpoint and petechial hemorrhages [597, 601].

In case of severe septic metritis in cow with fatal outcome, macroscopically the uterus was enlarged, highly edematous, and hemorrhagic. The cervical and endometrial mucosa was thickened and folded and contained copious amounts of discharge with dirty brown color. The myometrium is highly edematous. Process of inflammation was affected and fallopian tubes, where discovered areas of expansion filled with yellowish-brown exudate, alternating with obstruction of the lumen [597]. The liver and spleen of the cow had volume increased by approximately one-third. In the lungs were observed congestive hyperemia and moderate edema. It was also found acute swelling of a number of groups of lymph nodes. On serosa of the pericardium, spleen and liver were found small hemorrhages.

11.1.2 Microscopic Picture

Histopathological findings in the liver of infected sheep are characterized by perivascular lymphoid and histiocytic proliferations in the interstitium, single lymphocyte nodules in the parenchyma, granular dystrophy of the cytoplasm of hepatocytes, and hyperemia.

Picture of catarrhal lymphadenitis with reticuloendothelial hyperplasia is set to the mesenteric lymph nodes. In the study of a ewe-lamb were also observed single necro-biotic foci. Common finding was filling with blood vessel thrombotic masses.

In the lungs, in all cases observed activation of monocyte macrophage system. In two of the sheep was a picture of interstitial pneumonia. In the peribronchial and perivascular areas were identified mononuclear clusters of different sizes.

In the kidney, adrenal, heart, and spleen are established poor vascular and degenerative changes.

At the brain is set hyperemia with the activation of endothelial cells and in some cases perivascular mononuclear proliferations [621].

When carrying out microscopic examination of non-viable and stillborn lambs pathologic changes are found in almost all internal organs.

In the liver around blood vessels in the interstitium, we observed moderate lymphoid and histiocytic proliferations (Figure 11.1). At most cases, there are seen single lymphocyte nodules with necrotic changes in the center (Figure 11.2). There were moderate diffuse hyperemia and single hemorrhages [590].

The microscopic picture in the lungs is characterized by sealing the alveolar septa of certain areas as a result of multiplication of the intraseptal

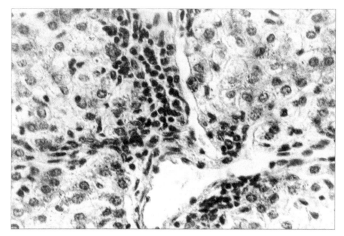

Figure 11.1   Experimental Q fever in sheep. Histological section of the liver of a stillborn lamb. Perivascular lymphocytic and histiocytic proliferation with single leukocytes. x250 /H & E stain/.

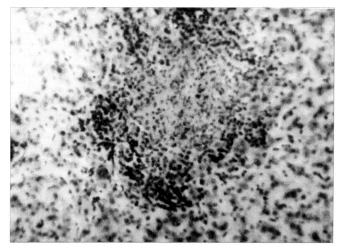

Figure 11.2   Liver of a stillborn lamb after experimental Q fever. Histological section. Lymphocyte nodule with necrobiosis in the center. x250 /H & E stain/.

cells and infiltration of lymphocytes (Figure 11.3). Degenerative and necrotic changes (lysis of the cell nucleus and desquamation) were observed in the epithelial cells of the bronchi. Around some vessels and bronchi were observed cuffs from mononuclear cells. In the large blood vessels establishes the activation of endothelial and adventitial cells filled with erythrocytes, while the smaller vessels have thrombosis [621].

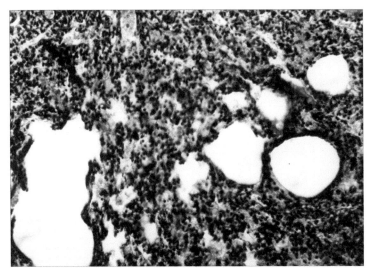

Figure 11.3 Lung of stillborn lamb after experimental Q fever. Histological section. Interstitial pneumonia. x160 /H & E stain/.

In the cortical portion of the kidney is found granular dystrophy, while in the medullary part vacuolar prevails. Mild lymphoid cell clumps are present in cortical and in medullary—hyperemia with the activation of endothelial cells.

In adrenal are prevalent degenerative changes (granular degeneration). In some sections of cortical noticed too weak lymphoid cell proliferation. There is hyperemia with the activation of endothelial cells, and the small blood vessels are affected by thrombosis.

Changes in heart muscle are characterized by a slight granular degeneration in myofibers, hyperemia, and activation of endothelial and adventitial cells.

In the spleen, we observed hyperplasia of the lymph nodes, while the reticular cells were highly activated.

In most cases, mesenteric lymph nodes are hyperemic and edematous and sinuses are filled with lymphocytes, leukocytes, and cells from plasmocyte order.

Sometimes simultaneously are observed increase of the cellular elements in the sinuses and diffuse increase in the number of lymphocytes and lymph follicles. Around some blood vessels are established small mononuclear proliferations, while in the small vessels thrombosis was detected (Figure 11.4).

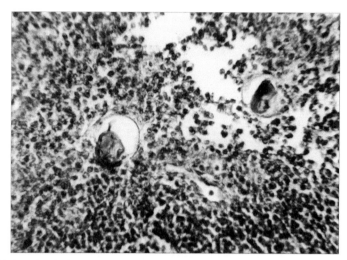

Figure 11.4   Histological sections of mesenteric lymph node of a stillborn lamb after experimentally induced Q fever in ewe. Thrombosis of blood vessels. x250 /H & E stain/.

The thymus in comparison with the healthy control animals indicates hyperemia, edema, and lymphocyte infiltration in the interstitium.

In the meninges are found hyperemia, edema, infiltration of lymphocytes, and single leukocytes. In the big brain is a strong pericellular and perivascular edema, imparting porous type of the cerebral substance. Adventitial and endothelial cells of the hyperemic blood vessels are activated as the smaller ones are surrounded by clusters of histiocytes and single leukocytes. Similar vascular and degenerative changes were observed in the medulla and spinal cord. The ganglion cells are usually lysed, and more rarely with pyknotic nuclei, such as around a few small nodules were detected (Figure 11.5).

The cerebellum showed ubiquitous edema with vacuolation of the brain tissue and perivascular and pericellular edema conferring loose structure, especially in the molecular layer. Strong edema is observed in the region of the Purkinje cells. In recent establishes lysis or pyknosis of nuclei. Similar lesions, but less pronounced, were found in the nuclei of the glial cells in the granular layer. Around some of them can be seen clusters of glial cells showing events of nevro–phagia. Other morphological features were diffuse hyperemia with single hemorrhages and thrombosis in part of the small blood vessels [621].

The fetal part of the placenta in some cases showed edema and desquamation of chorionic epithelium and leukocyte clusters. In other cases, the main

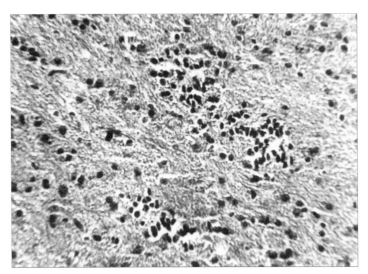

Figure 11.5 Experimental Q fever. Histological section. Big brain of a stillborn lamb. Glial nodular proliferation. x250 /H & E stain/.

findings were necrotic foci with lymph–leukocyte infiltration, thrombosis of some of the vessels, diffuse hyperemia, and bleeding. Necrotic–hemorrhagic changes were observed in chorionic epithelium and in the central parts of caruncles. Next to the caruncles slab are detected necrotic foci of different sizes with plenty of leukocyte infiltration and splits from C. burnetii cells (Figure 11.6). The uterine glands are enlarged, and some of them are filled with leukocytes. Myometrium is swollen with low leukocyte infiltration [590, 621]. Upon microscopic examination of impression preparations of placenta and uterus prepared by the immunofluorescence method, is detected C. burnetii, most commonly with intracellular location. Stained by Macchiavello [307], they are fine ruby-colored rods or polymorphic formations with the same location. Conducted electron microscopy of ultrathin sections of placentas revealed the presence of C. burnetii showing typical ultrastructural features (see Chapter 4).

The microscopic changes found in cases of respiratory disease in ruminants are mainly interstitial pneumonia, accompanied by a marked lymphocytic proliferation, particularly in areas around the bronchi, swelling and enlargement of the alveolar septa and regional lymphadenitis. At some cases there has been exudative pneumonia, pneumonia or bronchitis only and peribronchitis with varying degrees and extent of the damage [597].

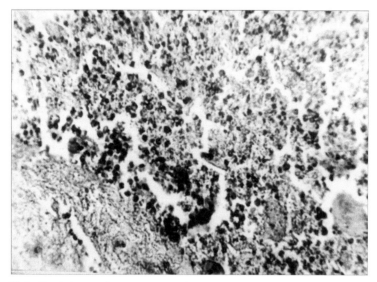

Figure 11.6 Fetal placenta from an eye experimentally infected with C. burnetii. Histological section. Necrotic focus with abundant subepithelial infiltration of leukocytes. x160 /H & E stain/.

11.2 Pathomorphological Investigations in Some Laboratory Animals with Etiologically Proven Q Fever

11.2.1 Macroscopic Picture

Rabbits. In experimentally infected intraperitoneal and intravenous route male rabbits, pathological changes were less pronounced. The liver and kidneys are hemorrhagic, and the spleen is rounded, almost twice increased. The mesenteric lymph nodes are hypertrophic and juicy. On the surface of hyperemic lung are established miliary, lightly densified foci with rose red color. Similar changes were observed in females miscarried after intraperitoneal and intravenous infection with C. burnetii, but they are manifested to a lesser degree. The uterus is enlarged and sometimes has red and swollen mucosa.

In aborted fetuses we found hyperemia and edema of the liver and spleen, and weak and off-white deposits on the surface. The changes in the placenta, unlike fetuses, are more pronounced. In almost all cases, it is swollen, hemorrhagic, thickened, and lost with transparency. The chorion is diffusely thickened, folded, and has a dirty red color. In some cases, on its surface are found whitish deposits [597, 622].

The macroscopic picture in necropsied rabbits infected intratesticular with C. burnetii isolated from a cow with severe metritis is characterized by two main morphological attributes: strong inflammatory edema with hypertrophy of testicles and the scrotal sack and splenomegaly [597]. In these two organs with distinct pathological changes is the main location of the infectious agent which finds without difficulty in examinations of impression preparations stained according to Stamp [487] or Macchiavello [307]. We watched too mild hepatomegaly.

Guinea pigs. In the preceding sections of this monograph, we indicated that intraperitoneal infection was the most common method of reproduction of Q fever in guinea pigs with a pathogen having different species and nosological origin. Induced diseases differed in clinical parameters of several indicators (see there). Pathomorphological changes also show variation with respect to the range and severity of lesions in various organs following infection with different starting infectious materials. When Q fever is reproduced with each of the used isolates, general macroscopic feature was enlargement of the spleen from two to five times established between the 16th and 24th day after infection. The color of the organ is dark red to purple, its density and consistency change and it easily breaks at moderate pressures. Upon necropsy of the guinea pigs on the fifth–sixth day after infection, we have found significant splenomegaly, but the increase of the spleen was less as compared with the above described in later periods. Hyperemia and edema of the kidneys, liver, stomach, and intestines were also observed. In some cases, on the surface of the spleens were detected off-white fibrin taxed. Frequent macroscopic findings are the tiny granulomatous formations on the outer surface of the spleen and in sections of its tissues [597].

The pathological changes in the spleen and other internal organs of guinea pigs after subcutaneous infection with C. burnetii of sheep origin are less pronounced. Splenic enlargement is more moderate, and changes in the liver, kidneys, and gastrointestinal tract are hardly visible or absent. The observed local skin reaction at the site of inoculation is anti-inflammatory process with hyperemia, swelling and small hemorrhages, penetrated deeply and affected the subcutaneous connective tissue and underlying muscle.

Intratesticular infection of guinea pigs with strains of C. burnetii having caprine, bovine, and human origin induced diseases in which macroscopically we found strongly expressed periorchitis and orchitis and enlargement of the spleen. The testicles were greatly increased by stretched red skin and blood-filled vessels. When cutting the testicles and scrotum reveals the features of diffuse hyperemia, inflammatory edema, swelling and tissue hypertrophy [597].

White mice. Macroscopically in the brains of mice after intracerebral infection with the agent detect different degrees of hyperemia of the meninges with slight edema and increase in the amount of the cerebrospinal fluid. Most blood vessels are dilated [597].

The most consistent pathological feature in intraperitoneal inoculation of the mice with different strains of C. burnetii is splenic enlargement. This increase can be several times larger compared to the normal size of the spleen. The latter is hyperemic, with rounded edges. When examining the microscopic preparations of external incisions and spleen surfaces were established rich deposits of the agent. Upon infection of the mice with the pathogen at high multiplicity and with more virulent strains, except lesions in the spleen, macroscopically were visible changes in the liver, gastrointestinal tract and kidneys: hyperemia and edema, and in the lungs – pneumonic foci of various sizes [597].

The reproduction of Q fever in white mice after intranasal infection was accompanied by distinct macroscopic inflammatory lesions in the lungs as a major morphological mark regularly discovered in autopsies between the seventh and the fourteenth day after inoculation. At the same time, the second regular finding registered an increase in the size of the spleen [597].

11.2.2 Microscopic Picture in Rabbits

In the liver of aborted rabbits in Glisson's triangles, around the blood vessels are found focal proliferation of lymphocytes, histiocytes, and single leukocytes. The proliferative processes are more pronounced at the 30th day from the inoculation in contrast to vascular and dystrophic processes which weaken. There are hyperemia, hemorrhage, and granular degeneration of the hepatocytes [622].

In the lungs were revealed lesions associated with various in severity interstitial pneumonia. Peribronchial are sometimes found large foci of mononuclear cells and single macrophages (Figure 11.7). There is hyperemia with hemorrhages and hyalinization of the walls of the alveoli. Microscopic picture of the lungs is characterized by swelling and thickening of the alveolar septa as a result of multiplication of the interstitial cells and the infiltration of lymphocytes. Some of the hyperemic blood vessels are surrounded by mononuclear proliferations [622].

In the kidney, adrenal glands, heart, and spleen are established weak vascular proliferative and dystrophic changes.

The changes in the lymph nodes are manifested with a picture of lymphadenitis, thrombosis of the smaller blood vessels, and partial hyalinization of the walls of some of them.

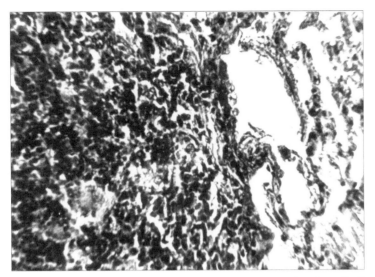

Figure 11.7 Experimental Q fever in a rabbit. Histological section of the lung. Peribronchial clusters of mononuclear cells and single macrophages. x250 /H & E stain/.

In meninges are situated edema, hyperemia, and infiltration of lymphocytes and leukocytes. In the cerebellum is established perivascular and pericellular edema, pronounced in the Purkinje cells. Hyperemia was observed with the activation of endothelial cells. Adventitial cells proliferate around some of the smaller vessels. Similar vascular and proliferative changes were found in the big brain, while in the medulla oblongata there are primarily vascular and dystrophic changes (of the ganglion cells).

In the liver of aborted fetuses is observed well-pronounced activation of the monocyte macrophage system, and in Glisson's triangles around blood vessels and bile ducts—perivascular lymphocyte histiocytic proliferations (Figure 11.8). In the parenchyma near the central vein is found focal and diffuse lymphocytic infiltration. The epithelial cells are vacuolated and have less granular cytoplasm, and in some places they are in a state of necrobiosis.

Poor lymphocytic and histiocytic clusters are located in the cortical portion of the kidney. There is hyperemia with a strong activation of the endothelial cells of the ducts of the glomeruli. The epithelial cells are grain dystrophic and completely lysed in some places.

In the cortex of adrenal glands are established degenerative and necrotic changes of epithelial cells and in the core—hyperemia with activation of endothelial cells.

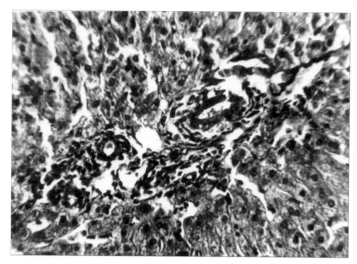

Figure 11.8   Experimental Q fever in a rabbit. Liver of the fetus. Histological section. Lymph histiocytic proliferation in Glisson's triangle. x250 /H & E stain/.

In the heart muscle is established the activation of endothelial and adventitial cells, sometimes distracted and focal lymphocytic infiltration (Figure 11.9).

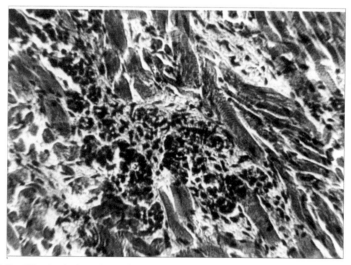

Figure 11.9   Histological sections of the heart of the fetus of rabbit after experimental infection with C. burnetii. Focal lymphocytic and leukocyte infiltration. x250 /H & E stain/.

The changes in the spleen are manifested by hyperemia and hyperplasia in the reticular and lymphoid tissue.

In the cerebrum of fetuses is established hyperemia with the activation of endothelial cells, as well as perivascular and pericellular edema. Around some of the larger blood vessels were detected mononuclear proliferations (Figure 11.10). The meninges are hyperemic, and occasionally, there is a lymphocyte and leukocyte infiltration. The ganglion and Purkinje cells are affected by degenerative processes [622].

In the fetal part of the placenta is established edema, desquamation of chorionic epithelium and lymphocytic and leukocyte infiltration (Figure 11.11). There are hyperemia and hemorrhage. In the maternal part of the fetal placenta occur more rarely necrotic and more often necrobiotic changes. In the endometrium of the uterus of animals aborted are established hyperemia and activation of uterine cellular elements. The myometrium is edematous with many mononuclear clusters around some of the blood vessels [622].

In light microscopy of impression smears from spleen, uterus and placenta from aborted rabbits, colored by Macchiavello [307], C. burnetii is seen most often in the form of intracellular fine ruby-colored rods or variously shaped formations.

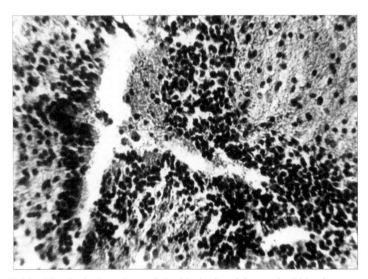

Figure 11.10 Histological sections of the big brain of fetus of rabbit after experimental infection with C. burnetii. Perivascular mononuclear proliferations. x250 /H & E stain/.

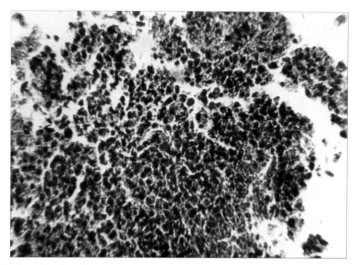

Figure 11.11   Abortion in a rabbit following experimental infection with C. burnetii. Histological section of fetal placenta. Focus of necrosis with leukocyte infiltration. x250 /H & E stain/.

Pathomorphological changes reflect the main peculiarity of Q fever, especially distinctly registered in experimental infection, namely the specific infectious fever that leads to vascular and degenerative changes in all parenchymal organs, pneumonia, lymphadenitis, proliferation, and inflammation of the placenta [590]. Comparison of the postmortem findings with the clinic of the disease can be seen that the observed clinical picture does not normally correspond to the intensity and localization of pathomorphological changes in the organism. In this regard, we share the opinion of Pandarov and Dimitrov [48].

Placentitis is one of the important pathological processes. In fetal and maternal parts of the placenta found degenerative, necrotic, and hemodynamic processes that we take for directional signs, but pathogenomic finding was the discovery of C. burnetii mostly in or around the necrotic tissue [590, 621]. Methods of electron microscopy complemented with pathohistological studies of placentas and allowed the determination of the concentration of C. burnetii in them. The placentitis induced by C. burnetii is different from the chlamydial placentitis in chlamydial abortion in sheep and goats, where are regularly seen more pronounced necrotic, membrane-like taxed and focal precipitation of calcium [590]. In Q fever except the placental tissue prejudice massively cotyledons, where develop inflammatory-necrotic processes [621]. For similar findings, but with different shades of histological features such as integrated components of the overall microscopic picture, reports and other authors [48, 286, 294, 360, 415, 456, 581].

The presented results from pathomorphological research on stillbirths or deliveries of weak non-viable lambs showed that Q fever occurs as a generalized infection [621]. All internal organs are affected in varying degrees from inflammatory processes. The changes are diffuse. The most pronounced is the damage to the liver, lungs, brain, and fetal placenta. Frequent findings were the lymphocytic and histiocytic proliferations around some inter-lobular veins lymphocyte clusters in the hepatic parenchyma and necrobiosis of hepatocytes. The inflammatory changes in the lymph nodes (lymphadenitis), the lungs (start of interstitial pneumonia), and the brain (encephalitis) are also frequently observed. The changes described in the internal organs and brain are similar to those established by other researchers [10, 291, 559] in lambs and sheep, but in our studies of stillborn and weak lambs, the proliferative, degenerate, and vascular changes are more pronounced.

Regarding the localization of the processes, changes in the internal parenchymal organs, brain and placenta in experimentally-induced infection are correlated with the established lesions in aborted fetuses in spontaneous Q fever. Concerning the power of expression, particularly the vascular changes (thrombosis of blood vessels) and the proliferative changes in the lungs, lymph nodes and brain are more pronounced in the experimental infection of non-viable and stillborn lambs [621].

Indicative of experimentally infected with C. burnetii sheep are the perivascular lymphocytic and histiocytic proliferations in the interstitium of the liver, the interstitial pneumonia and lymphadenitis with thrombosis of blood vessels [621]. Similar findings were reported by other authors [48].

Pathomorphological changes in guinea pigs and white mice in unison with the clinical manifestations showed variations with respect to the location, nature, extent, and severity of lesions in various organs after infection with C. burnetii of different types and nosological origin. Too characteristic of the most commonly used method of intraperitoneal infection in order to reproduce the Q fever is the development of generalized infection accompanied by inflammatory and degenerative changes in the liver and spleen [597]. These findings are in agreement with the data of other researchers [18, 48, 52]. Splenic enlargement—most pronounced in i.p. inoculation—was also regularly present in most other ways to inoculate the pathogen, where stood out other pathomorphological changes: pneumonia (nasal); orchitis and periorchitis with diffuse hyperemia, inflammatory edema, and tissue hypertrophy (intratesticular); and local inflammatory process of the skin, the subcutaneous connective tissue, and the adjacent muscles (s.c.).

Compared with guinea pigs and mice, rabbits are used less frequently in experimental work. Therefore the information about various aspects of studies on infection with C.burnetii in rabbits, including clinical and pathological findings is insufficient. The results of our research suggest that the infected i.p. rabbits develop infectious process leading to fever, abortion, accumulation of specific antibodies, and inflammatory and degenerative lesions in parenchymal organs, lymph nodes, and brain [622]. These data are similar to the observations in other laboratory models—guinea pigs and mice [11, 18, 32, 48]. In infected rabbits, the most affected were the liver, lungs, and lymph nodes. It is noted that degenerative and vascular changes occur earlier, while proliferative are best manifested in 30th day after infection. Around and in the walls of the interlobular hepatic veins are established focal congregations of mononuclear cells and in the liver parenchyma—lymphocyte–leukocyte nodules and necrobiosis of the hepatocytes. The lungs are affected by interstitial pneumonia. In the lymph nodes, there is a lymphadenitis with thrombosis of the small vessels [622]. Most of the described changes are observed in other laboratory animals [18, 48]. Comparing the morphological changes in rabbits infected with sheep strain of C. burnetii and those infected with blood (from ewe-lambs in bacteremia of C. burnetii), we conclude that in terms of localization, the processes coincide, but in order of appearance are more pronounced following infection with the strain [622].

In the literature, we found no evidence of postmortem studies of aborted fetuses and placentas from rabbits with Q fever. Our research in this area shows that disease agent passes the placental barrier and causes abortion and morphological lesions in the fetuses. The established proliferative and dystrophic changes in the liver, lungs, kidneys, spleen, and brain of fetuses of rabbits are similar in nature to those observed changes in the relevant organs of ovine fetuses infected with C. burnetii [621, 622]. The histological changes in the liver and the rest of fetal parenchymal organs were observed 10 days after infection, but were best manifested to 20 days. This gives us reason to assume that like the sheep and rabbits, inoculating of the agent should be done in the second half of pregnancy. The observed clinical signs and all the morphological changes show that the rabbits can be used successfully for the diagnosis and research of Q fever. The acquired knowledge of abortions in the rabbits induced by C. burnetii under experimental conditions may be useful in the study of miscarriages in these animals on farms or in private yards, which would lead to epizootiologic, epidemiological, and economic impact.

12

Epizootiology and Epidemiology

Q fever is a zoonotic disease caused by extremely infectious agent affecting a wide range of hosts, forming two types of foci of the infection—natural and agricultural. Reservoir of the infection in nature is ticks, rodents, wild ungulates, and wild birds. There are over 70 species of ticks, mainly ixodic, where it is established spontaneous infection with C. burnetii [22, 30, 85, 405, 515]. In many species soon after infecting them it appears transient bacteremia with C. burnetii—a period in which thriving on them ticks can be infected at the bloodsucking [343]. The agent multiplies in the middle part of the intestinal tract or stomach of the tick and is excreted in large quantities with its excrements, which fall onto the skin of the animal. In some ticks, C. burnetii multiply in all organs and besides with feces are released with the saliva [46]. It has been proven transmission of C. burnetii of the offspring through eggs (trans-ovarian), which is an essential factor for the persistence of the infection in tick populations [46, 85]. Coxiella burnetii in the ticks is in a first phase and has high infectivity. Circulating in the external environment, the agent from the ticks infects through transmissible route (by biting and sucking), aerosol route, and alimentary route the wild mammals and birds and the domestic animals when they are grazing [85, 177, 318, 465, 491, 522, 525]. The infected rodents and other wild mammals and birds also release the agent with their secretions and excretions, and serve as a reservoir and source of infection for other animals and ticks. Thus, in nature extensive active foci of the infection are maintained with C. burnetii with a high degree of virulence of the pathogen for domestic mammals, birds, and also people by contact with them [46, 85].

Farm animals in contacts with the natural foci become infected and form massive agricultural reservoir of C. burnetii, which is a source of spread of infection to other animals and humans. Humans have much less frequent contacts with the natural foci, and therefore, the agricultural foci play a key role in the epidemiology of Q fever for people [46, 69, 305, 353, 509, 564, 597]. Rehaček et al. [431] established serologically wide dissemination of

C. burnetii in cattle in southern Bavaria, Germany, and presuppose the existence of independent natural cycle of the agent in this species. The infected domestic ruminants (cattle, sheep, goats) release C. burnetii with birth products, feces, urine, and milk from which other animals and humans can become infected. Particularly strong excretion and disseminating the infection are through placentas and amniotic fluid during births or in the case of abortion [44, 456, 509, 525]. Further, the infections are performed by respiratory or alimentary route. Epizootiological data of some authors have shown that the cows more often than sheep develop chronic infection with C. burnetii and longer release the pathogen in the milk, which defines them as the most important source of infection for humans [85, 111, 288, 343]. According to Lang [288], goats, like dairy cows, show predisposition to develop chronic Q fever. The significant increase in the number of goats reared in a given country represents a risk factor for human infection with C. burnetii in contact with them [288, 447, 471, 594]. Goats play a key role as a reservoir and source of infection in the largest ever known epidemic of Q fever in the Netherlands (2007–2010) with a total registered 4026 human cases [603, 623, 624].

Other species of domestic mammals (horses, pigs, buffaloes, camels, rabbits, dogs, cats) and birds (ducks, geese, chickens, turkeys, pigeons) can also be infected with C. burnetii under natural conditions with different ranges of morbidity [242, 252, 269, 338, 349, 522, 606].

There is also the view that each species can be infected with C. burnetii, but the main risk for infection of humans comes from cattle, sheep, and goats [525].

For dogs especially in rural outbreaks of Q fever had long data that are susceptible to C. burnetii [314]. The infection to them was carried out in an alimentary route (eating placentas, milk, and other foods contaminated with the agent from diseased ruminants), inhalation of infectious aerosols, and dust and transmissible by parasitic on their skin ticks. Dogs affected by Q fever can be a source of infection for humans [115, 290, 429]. Laughlin et al. [290] describe the epidemiological chain of infection with C. burnetii: deer–dog–man.

The last decade has increased the interest to Q fever in cats, after describing diseases in humans, which have been developed after contact with birthing animals of this species [242, 289, 319, 338, 399, 606]. Komiya et al. [278] investigated the prevalence of C. burnetii in cats in Japan living in different environments and establish three times higher seroprevalence in stray cats (41.7%) compared to animals reared at home (14.2%). British authors [546] suggest that an important reservoir of C. burnetii, of which domestic animals, especially the cats become infected, is the wild brown rat (Rattus norvegicus).

Several cases of human illness are described due to direct or indirect contact with the sources of infection. Even in the initial period of the studies of the disease, the importance of Q fever in troops is highlighted, due to the specific conditions of work and life [3, 118]. There is documented information about thousands of cases of diseases of military servicemen during World War II and in later periods associated with the stay of troops in areas where there are, or were animals infected with the agent of Q fever (farms, corresponding pastures, and others). These are described in earlier publications of epidemics, outbreaks, and sporadic cases in Greece, Serbia, Italy, Corsica, Crimea and Ukraine (1944–1945), Libya (1951), Algeria (1955), Isle of Man (1958), which are summarized in a review of Spicer, 1978 [483]. The likely route of infection of the military persons is respiratory, and in some cases alimentary [483]. Later (1991–1994), acute diseases of Q fever have been identified among American servicemen during the Gulf War—meningoencephalitis [190], pneumonia [292], and acute seroconversion [541]. Epidemiological information in those cases shows that diseased persons have been in contact with sheep, goats, and camels. In the same period, Q fever is identified in US soldiers in Somalia and among Somali refugees suffering from acute febrile illness of unknown origin [213, 310, 541]. Cekanač et al. [495] described epidemic outbreak among 20 soldiers of the Yugoslav Army during the war, after the aerial powder contamination with C. burnetii of sheep. Serologically proven Q fever in patients clinically has been manifested by pneumonia (35%) or a flu-like condition. The presence of C. burnetii in the urine, feces, and birth products of the infected animals, the contamination of the external environment, including the implementation immediately located to the animals hay, straw, clothing, and supplies, favors either direct infection of humans by inhalation of infectious aerosols or dust, as the transfer of the agent to individuals who are not in direct contact with the infected animals [81, 90, 139, 323, 469].

Other endangered groups in terms of Q fever are livestock and veterinary staff, workers in slaughterhouses, dairy farms and enterprises processing wool, personalized service laundries handling contaminated clothing, laboratory workers, residents of settlements in routes by which pass trucks serving farms [118, 169, 212, 224, 517, 530, 564].

In the majority of publications on the epidemiology of Q fever, respiratory route of infection is seen as essential in humans [427].

The big resistance of C. burnetii in the external environment favors the storage and distribution of microorganism. There are reports of transmission of the pathogen of agricultural foci Q fever by the wind [305, 328, 519, 597].

The dry weather favors the formation of infectious dusts and aerosols [305]. Except through the natural wind created due to human activities, drafts can also transmit C. burnetii with further cases of infection [81, 533]. Armengaud et al. [81] describe epidemic outbreak among 29 citizens of the French city Brianchon (12,000 inhabitants) who are not engaged in professional risk classes. It is believed that infection has been through respiratory due to dust from the air continuously in helicopters' landing and taking off next to the meat plant infected during the slaughter of sheep and goats [81]. Van Woerden et al. 2004 [533] investigated epidemic with 282 patients in Wales, UK, occurred in industrial enterprise during the renovation of an old building with thatched walls and ceilings. In the destruction of the latter is formed cloud of dust containing large amounts of spore-like forms of C. burnetii, which extends to the neighboring workplaces and is inhaled by the staff [533].

The consumption of infected raw cow and goat milk and goat cheese also can lead to diseases in the people [93, 194, 285, 355, 509, 625]. Upon epidemiological studies in Canada of an epidemic of Q fever associated with goats, Hatchette et al. [228] allow connection between the consumption of cheese from pasteurized goat's milk and certain part (17/60/28.3%) of infected persons, which was reported for the first time in the literature. This assumption needs further studies, since C. burnetii has not been isolated from the cheese [216].

Described are very rare cases of human infection by other means: when contacts by obstetrician with an infected woman during an abortion performed by him [423] and by transplacental road leading to the death of the fetus and abortion [494]. In earlier publications (Herman, 1949; Marmion, Stoker, 1950), contamination of doctors during autopsies is reported. Single cases of infection from tick bites in nature are described by Eklund et al. [178] and Beaman and Hung 1989 [91].

In the men the infection with C. burnetii occurs more frequently than among females—approximately in the ratio 1.5–3.5: 1, according to Migala [355] and according to Tissot Dupont [517]—2.5: 1. These ratios are contingent because they are directly dependent on the more frequent involvement of men in professions that are at risk of infection with Q fever [355]. The average age of the infected individuals is approximately 45–50 [355] and, according to another source [343]—30–60 years, coinciding with the active employment of the population in this age range. The spread of Q fever among children was probably more than was thought until recently and obviously requires targeted research—Maltesou et al. [313]; Kazar 1999 [273]; Ruiz-Contreras et al. [449].

For the issue of seasonality in the incidence of Q fever in animals and humans there is no single answer because of various data obtained in different countries and regions and difficulties in accurately registering the cases as a result of often non-specific or absent clinical manifestations [343]. Overall, in sheep and goats, it is considered that the highest number of infections is in winter and spring in connection with the mass births, abortions, and hard lactation and also in the summer, in the period of greatest biological activity of ticks and wild rodents [46]. In cattle is not apparent seasonality of the disease. In all cases, the structure of livestock production practices in the farming and other organizational factors affects the level of permeates and seasonality in the manifestation of the disease, and this reflects on the morbidity in humans. In Europe, the cases of acute Q fever are more frequently observed in the spring and early summer [517].

In the contemporary conditions are observed a number of deviations from the classical epidemiology of Q fever related to seasonality, risk groups, age ranges and more—Hellenbrand, [241]; Raoult et al. [426].

The geographical spread of Q fever has a global character [126, 156, 188, 196, 331, 355, 510, 533, 626]. Based on existing information, it is considered that only New Zealand is free of the disease [214, 243]. The prevalence of the disease among domestic animals was found in several earlier publications mainly in the 1960s of the twentieth century, some of which are cited in the review of the historical data of this work. Nowadays, the actual presence and distribution of C. burnetii are based on modern sero-epidemiological studies. Undoubtedly of great interest is the dynamics of the spread of Q fever in animals and related diseases in humans. For example, Lang [288] and Krauss [283] indicate that the incidence of cattle from Q fever is higher than 20–30 years ago. Extensive research and analysis of Q fever in Germany by Hellebrand et al. [241] covering retrospective period from 1947 to 1999 show that C. burnetii infection is endemic among cattle throughout the country and sheep in the southern, eastern, and western parts of Germany. The diseases in humans have increased in the last decade of the study, and the main source of infection for humans are the sheep. It concludes that conducted structural changes in the livestock production in the 1990s led to an increased number of outbreaks of Q fever in humans [241, 306, 433]. Large modern studies in the USA—Jay-Russell et al. [264], McQuiston et al. [349]—prove the prevalence of infection with C. burnetii in animals and humans in almost every state. In many other countries and regions has been established the contemporary status of Q fever in animals and humans. In some of these studies, the positive findings were first detection of the disease: Vojvodina—Republic of Serbia

[468]—sheep, people; Austria [274, 275]—sheep, cattle; Czech Republic and Slovakia [169, 302, 303]—sheep, cattle; Poland [78, 79]—cattle, sheep, bisons, people; Hungary [294]—sheep; France [81, 343, 519]—sheep, goats, people; UK [514, 533, 564]—sheep, cattle, poultry, people; Spain [424, 459]— sheep, goats, cattle, people; Italy [336, 388, 469]—cattle, sheep, people; Switzerland [172, 173]—sheep, people; Greece [313, 486, 524]—sheep, goats, people; Bosnia and Herzegovina [350]—people; Turkey [576]—cattle; Russia [447]—cattle, sheep, goats, people; Israel [477, 570]—humans, cattle, sheep; the Netherlands [603–605, 623, 624]—goats, sheep, people. Other publications presented the evidence of nosological geography of Q fever in Belgium, Denmark, Ireland, and Portugal [185, 251]. Scarcer is the data on the spread of disease in Africa, where there is clearly insufficient diagnosis of the disease. Kolb [279] establishes C. burnetii infection in outbreaks of abortions in goats in Namibia and Rohde et al. [441] demonstrated an considerable range of the disease among dairy cows in Zimbabwe. Q fever among cats and related diseases in humans was recorded in South Africa and Zimbabwe [338, 339] and in Niger in children under 5 years, especially in the case of close contact with domestic animals [265]. Serological surveys have revealed antibodies against C. burnetii in humans in Nigeria—44% [101] and 10–37 percent in Northeast Africa [198]. In Jordan, Q fever has been found in sheep and goats [71]. The disease is also established in Oman (people), Saudi Arabia (humans and domestic ruminants), and Abu Dhabi (UAE)—in racing camels [467]. Studies in Japan show that Q fever has enzootic character among animals— cattle, sheep, goats, cats, dogs, birds [252, 278, 364, 373, 522, 568, 569] and a significant spread in children with atypical pneumonia and influenza—like symptoms [372, 521]. In South Korea, C. burnetii infection was found in cattle [267, 268] and in people with pneumonia, hepatomegaly, and splenomegaly [135]. In Thailand, the highest seropositivity have the dogs. The disease is also detected in goats, sheep, cattle, and humans [505]. Data on the prevalence of Q fever in Canada are related to the goats, sheep, dogs, cats, hares, and deer [228, 290, 456, 457, 530]. For diseases caused by C. burnetii in northern Peru (Andes) reported Blair et al. 2004 [100].

12.1 State of the Natural Foci of Q Fever in Bulgaria

12.1.1 Ticks

To clarify the status of the natural foci (NF) in 1993–2004 studied for carriage of C. burnetii, 1769 ixodic ticks were collected from domestic animals in different ecological and geographical areas of the country [606, 627].

By using immunofluorescent hemocytic test (IFHT), it was found that infected with C. burnetii were 298 (16.84%).

The study performed in two stages. The first covered the 8-year period (1993–2001) and then examined 846 ticks or an average of 105 per year. The ticks were collected from domestic ruminants in 18 regions belonging to the species Dermacentor marginatus, Rhipicephalus bursa, and Hyalomma plumbeum. Positive for C. burnetii were 135 (15.95%).

The second stage of the study was a three year (2002–2004) and provided more recent information on the tick reservoir of infection in the country. A total of 923 ticks were studied from 41 settlements in 13 districts: 306 in 2002, 309 in 2003, and 308 in 2004. Ticks collected from cattle were 310; goats—291; sheep—281 and dogs—41. The object of the study was nine species of ticks. Eight of them establish infection with the C. burnetii. The results showed that infected with the agent of Q fever are 163 copies (17.65%) from 36 villages in 13 districts. Positive results were obtained from a total of 52 catches of ticks. In another 30 catches, the agent was not found. The infestation of the ticks ranged from 8.33 to 55%.

When collected from cattle ticks 64 were positive for C. burnetii (20.64%). In sheep, goats, and dogs, the figures were respectively 51 (18.14%) 40 (13.74%), and 8 (19.51%).

Summed data by species ticks revealed the highest level of infection in Ixodes ricinus—46.34%, followed by Rhipicephalus sanguineus—27.65%, R. bursa—24.19%, Boophilus calcaratus—22.22%, D. marginatus—22.15%, H. plumbeum—21.25%, Haemaphysalis punctata—11.11%, and D. pictus—9.74%.

As frequency of detection of infected ticks in different caught, despite the low percentage of the positive first is the type R. bursa (16 catches), followed by D. marginatus—12, and H. plumbeum—9. Lowest frequency (1 catch) is registered at B. calcaratus and H. punctata.

In southwestern Bulgaria, carrier state of C. burnetii was found in D. pictus, R. bursa, and H. plumbeum, collected from sheep and goats. In northwestern Bulgaria, we observed infection with the agent primarily on D. marginatus from sheep, goats and cattle and to a lesser degree of I. ricinus and H. punctata cattle. In southern Bulgaria was found infection with C. burnetii in the species R. bursa (sheep and goats), R. sanguineus (goats, dogs and sheep), D. marginatus (cattle and sheep), and H. plumbeum (of cattle). In southeastern Bulgaria were infected R. bursa (cattle and sheep) and H. plumbeum (sheep) and in northeastern Bulgaria I. ricinus, D. marginatus, B. calcaratus, and R. bursa (all from cattle).

Figures 12.1–12.5 demonstrate the direct immunofluorescent hemocytic test applied in the examination of hemolympha of three types ixodic ticks: I. ricinus, D. marginatus, and R. bursa. Concern is for highly expressed positive

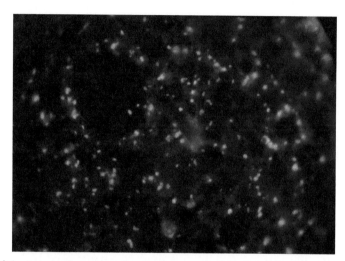

Figure 12.1   Immunofluorescent hemocytic test. Coxiella burnetii in hemolymph of an ixodic tick Ixodes ricinus. Staining by direct immunofluorescent method (IFM). Magnification 90x; 10x.

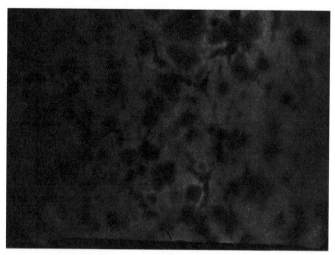

Figure 12.2   Immunofluorescent hemocytic test. A tick *Ixodes ricinus*—negative for infection with the C. burnetii. Staining by direct IFM. Magnification 90x; 10x.

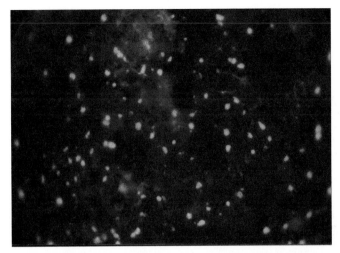

Figure 12.3 Immunofluorescent hemocytic test. Coxiella burnetii in hemolymph of an ixodic tick *Dermacentor marginatus*. Staining by direct IFM. Magnification 90x, 10x.

Figure 12.4 Immunofluorescent hemocytic test. A tick *Dermacentor marginatus*—negative for infection with the C. burnetii. Staining by direct IFM. Magnification 90x, 10x.

reactions in direct detection of C. burnetii (Figures 12.1, 12.3 and 12.5). In Figures 12.2 and 12.4 are shown negative results for C. burnetii.

From ticks were received nine isolates C. burnetii—seven in YS of CE (D. marginatus, R. bursa, R. sanguineus, H. plumbeum) and two in guinea pigs (D. marginatus, R. bursa).

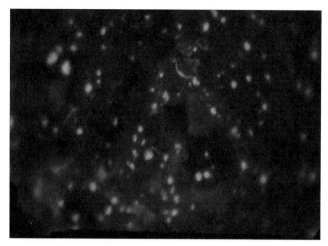

Figure 12.5   Immunofluorescent hemocytic test. Coxiella burnetii in hemolymph of a tick Rhipicephalus bursa. Staining by direct IFM. Magnification 90x, 10x.

The results show that there is some increasing infestation of ticks in the last three years—2002–2004, (1.7%) compared with the preceding period from 1993 to 2001, where the carriage of C. burnetii is 15.95% [606, 627].

12.2 Wild Mammals and Birds

The spread of Q fever among some wild mammals and birds was examined by serological and virological investigations of blood samples and clinical and pathological materials obtained from different regions of the country: Vd, Vr, Pl, SF Pk, VT, Rz, Ss, Hs, Kj, Sm, Pz, the Rhodopes, and the Rila pan.

Serologically were examined a total 990 blood samples from 17 animal species (Table 12.1). In 15 of them (except wild pigs and hamsters) were found positive seroreagents originating from almost all the regions [628]. Some of the investigated species are known as part of the natural reservoir focal of Q fever in Bulgaria—hamsters, crocidura (Sorex), hares, deer, magpies. In our studies, highest prevalence in this group showed crocidura—36.84%, and deer—32.23%.

Approximately the same range of infection was established in wild rabbits (25.8%) and magpies (26.22%). Surveyed 71 specimens European ground squirrel (Citellus citellus) were seropositive at 15.49%.

Antibodies against C. burnetii found was in five cases (8.92%) in the study of 56 foxes.

Table 12.1 Serological testing for Q fever of wild mammals and birds

Species	CFT Tested (No.)	Positive No.	Positive (%)
Deer	121	39	32.23
Mouflon	117	11	9.4
Wild goats	5	5	100.00
Hares	62	16	25.8
Foxes	56	5	8.92
Wild pigs	137	0	0
Hamsters	5	0	0
Ground squirrels	71	11	15.49
Field mouse	64	13	20.31
Forest voles	18	10	55.55
Yellow-necked mouse	11	4	36.36
Forest dormouse	7	3	42.85
Crocidura	19	7	36.84
Magpies	122	32	26.22
Doves	152	18	11.84
Pheasants	12	4	33.33
Ravens	11	3	27.27

The majority of the other wild species (Table 12.1) have not been the subject of serological tests in the country. We will emphasize that for the first time were obtained serological evidence of Q fever in mouflon (9.4%) and in wild rodent forest dormouse (Dyromys nitedulla)—42.85%. High rates of serologically positive for C. burnetii reactions we registered in other rodents—forest voles (Cletrionomys glareolus)—55.55%, yellow-necked mouse—(Apodemus flavicollis)—36.36%, and field mice—20.31%.

Surveyed five wild goats of game-breeding stations in Devin (Sm) and Velingrad (Pz) were seropositive for Q fever with CF titers (phase II) from 1:64 to 1:128. Despite the limited number of samples, these first studies of wild goats in the country are testified to substantial susceptibility of this species to infection with C. burnetii. In the birds, besides the above result of magpies stand serological findings in the pheasants (33.33%) and the ravens (27.27%), while specific antibodies in pigeons were detected in 11.84% (Table 12.1) [628].

The results of virological tests on clinical and pathological material from wild mammals and birds are exposed in Table 12.2. This shows that of seven types: wild rabbit, forest dormouse, pigeons, doves, crows, pheasants, and raven were obtained 12 isolates of C. burnetii—9 in chicken embryos, 1 in the brains of newborn white mice (NBM) and 1 in guinea pigs after

Table 12.2 Isolation and direct indication of Coxiella burnetii from wild mammals and birds

Species	Ecological and Geographical Area	Starting Clinical or Pathological Material	Demonstration of C. burnetii	
			Isolation of the Agent /No. of Isolates/	Direct Detection of the Agent Method
Wild rabbit	Hs	Blood	In the brain of NBM (1)	Direct EM
Forest dormouse	Sm	Organ suspension from spleen, liver and kidneys (pooled 3 copies)	CE (1)	Direct EM, LM
D. Nitedulla				
Pigeons	Vd	Spleen, liver, intestinal content	CE (2), guinea pigs—i.p. (1)	LM
Doves	Pl	Liver, lung, heart (pooled samples)	CE (2)	LM
Crows	Pl	Liver, lung, heart (pooled samples)	CE (2)	Direct EM, LM
Raven	Pl	Spleen, liver, lung	CE (1)	
Pheasants	Rz	Spleen, liver, lung (pooled samples)	CE (2)	Direct EM, LM

intraperitoneal inoculations. Reflected it is also the direct indication of the causative organism in EM and light microscopy smears of the output infectious materials (Table 12.2).

The results indicate the role of these wild animals as natural reservoirs and vectors of C. burnetii, having a certain share in the circulation of the infectious agent and in maintaining the natural and agricultural foci of Q fever.

12.3 Agricultural Foci of Q Fever in Bulgaria

12.3.1 General Characteristics

In the eighth chapter, data were presented for the disclosure of the agricultural foci (AF) of Q fever in Bulgaria, based on systematic and very large volume serology in the period 1977–2006.

The state of the agricultural foci based on data for the seven year period from 2000 to 2006 was characterized by the presence of 393 foci of Q fever among domestic ruminants as a major reservoir of C. burnetii, at that including 192 in cattle, 121 in sheep, and 80 in goats [620]. Isolations of the pathogen and its direct detection in different clinical conditions confirm the etiological nature and complement the epizootiological characteristics of Q fever and its nosologic geography in the country represented by numerous agricultural and natural foci [597, 598].

At conducted extensive research throughout the country, the establishment of new agricultural foci of Q fever was carried out by prophylactic or targeted serological and virological screening tests in clinical disorders combined with clinical observations and information from the epizootiological surveys. An essential element in these activities has been monitoring of the status, activity, and the dynamics of already known foci of Q fever in the relevant animal areas.

The agricultural foci of Q fever, similar to the natural foci, are *complex, changing multicomponent systems with a range of peculiarities and relationships* [597, 598, 620]. Depending on the size of the herds or groups of animals, respectively, the number of affected ruminants subdivides the foci in large, medium-sized, and small. At this differentiation with regard to the ways of raising animals and organizational structure of livestock farming, compact farms of different sizes or multiple small farms include single ruminants reared in the yards of the homes. We are taking into account the location of the farm— within the settlement outside, but near, or distant from settlements holdings, including summer camps for animals.

Table 12.3   Contemporary state of the agricultural foci of Q fever in Bulgaria

Foci of Infection with C. burnetii; Serologically Positive Animals	Economic and Epidemiological Significance
• 393 AF among cattle, Sheep, and goats.	Significant economic losses. Key role in the epidemiology of Q fever in humans.
• Single smaller foci associated with buffaloes (15), chickens (5), and ducks (2)	Potential risk of more widespread infection
• Single seropositive pigs and horses	Need for systematic serological screening and clinical and epizootiological observations

In many cases, we found foci of Q fever in ruminants that were close to each other within a particular region or area. This formed whole of large areas of massive and most often durable agricultural foci. In such foci of infection and conglomerates of them, we found significant percentage of serologically positive animals and the discovery of new seroreagents continued for years. Typical examples of this are the regions of Vratsa (outbreaks in Kneja and other places), Kyustendil, Pernik, some areas of Sofia district (Etropole, Botevgrad, and surrounding villages), and others.

The diseases of the main types of ruminants in formed large and medium agricultural foci have important economic and epidemiological significance, and the single minor foci and single seroreagents among other animal species determine the possible risks of escalation and disseminating infection among these populations (Table 12.3).

12.4 Internal Structure of the Agricultural Foci

Different internal structures of the foci of Q fever are discovered among domestic ruminants. Some of them, mainly in permanent flocks of sheep, goats and cows—owned by one person or company, are compact masses of animals in close daily contact with each other in stalls, pens, or pasture where the infection with C. burnetii spread mainly between them. The contacts of these animals with humans are primarily to the servicing them livestock and veterinary personnel. Any other type of herds—pooled from single-ruminant residents in private yards, but collected every day for grazing are too numerous in the last 10–12 years after the liquidation of large-scale social farming and its replacement with small fragmented private farms. In these herds, the infection with C. burnetii is transmitted at the daily collection of animals and their contacts on the roads and grazing. The infected ruminants are in a daily close contact with their owners at homes where the births take place and performed milking.

12.5 Categories of Animals in the Agricultural Foci

In the AF of Q fever, we find different categories of animals:

- clinically ill with the most active excretion of C. burnetii;
- subclinical, oligo-symptomàtic, and asymptomatic seropositive animals with different levels of excretion of the pathogen, but most often less as compared to the previous category;
- contact clinically healthy animals present in the incubation period;
- suffered without excretion of the agent;
- contact healthy animals.

The ratios of these categories were varied, depending on the intensity of the epizootic process, the virulence of circulating strains, the size of herds and types of farming, the effectiveness of the measures taken for limiting the disease, proximity to natural foci, and reinfection with C. burnetii derived from them.

12.6 Etiologically Proven Clinical Forms of Q Fever in the AF

Our comprehensive studies based on large etiological, clinical, and experimental material [590, 591, 594, 595, 597, 601, 602, 615, 616] allowed us to make the following *classification* of the clinically apparent conditions in Q fever in domestic ruminants:

- Abortions and related conditions of the pathology of pregnancy (premature births, stillbirths, and births of non-viable offspring);
- Respiratory diseases (pneumonia, bronchitis, influenza-like form);
- Mastitis;
- Endometritis and metritis;
- Reproductive disorders;
- Syndrome of prolonged reduction in milk production;
- Inapparent or latent form.

12.7 Autonomous Circulation Circle and Cyclical Reproduction of the Infection with C. burnetii in the Agricultural Foci

Our research shows that once created the agricultural foci of Q fever maintain separate (autonomous) circulation circle of C. burnetii in the animals, based on the carriage and excretion of the pathogen with the secretions and excretions and direct infection from animal to animal, without intermediate hosts and vectors [597, 620]. The activity of the foci of Q fever was maintained by constantly infecting new animals and making them clinically ill or asymptomatic infection carriers and new emitters of the infectious agent. This leads to a cyclical reproduction of the infection with C. burnetii.

On farms and herds affected by Q fever with epizootic character and clinical manifestations, the activity of the AF is determined by the massive exretion of the agent is greater. Such highly active foci of the disease were detected in the described outbreaks of miscarriages and related conditions in sheep, goats and cows, mass mastitis in sheep, and respiratory and genital infections.

In the cases where in herds, there was a Q fever with a significant percentage of seropositive ruminants, but with a small number of clinical diseases, focusing on the birth periods in sheep and goats, which were respectively the autumn–winter and winter–spring seasonal and year-round ongoing births in cows. In these periods was established shedding of the agent not only in cases of abortions, premature births, stillbirths, or unviable offsprings, but in normal births, accompanied by activation of the latent infection. Infectious agent is found in the amniotic fluid, fetal membranes, puerperal genital secretions, milk, urine, and feces.

From the main sources of the infection in AF—infected cattle, sheep and goats, the pathogen can be transmitted to the susceptible ruminants via alimentary and inhalation ways.

An important factor favoring the spread of Q fever was the uncontrolled movement of animals from infected areas and farms to disease-free zones and farms and vice versa. Particularly susceptible to infection were healthy ruminants delivered from other regions and introduced for the first time in infected flocks. Such animals most often develop the disease manifested by acute clinical signs and become active emitters of C. burnetii. On the other hand, the movement of contaminated animals in uninfected herds led to the introduction of the infection in the past and the emergence of new agricultural foci of Q fever. The disease was also found in a new assembled herds of sheep and goats purchased from different farms and owners, without prior serological tests for antibodies against C. burnetii. For example, in recent years in the region of Haskovo was revealed epizootic of Q fever in sheep in a new holding, wherein for short terms were collected about 500 animals. The massive disease, clinically manifested by abortions, stillbirths, and respiratory disorders, etiologically was proved by the discovery of 228 serologically positive animals and also through direct indication of the causative agent. In a similar survey in the region of Veliko Tarnovo, we found that to basic herd of 500 sheep of local breed are joined 200 newly purchased in another area Merino sheep, which led to epizootic outbreak of Q fever with 71 registered abortions to testing and specific seropositivity in 85.7% [597].

As a link in the epidemiological chain of Q fever should classify dogs living in farms and accompanying flocks of sheep, goats and cattle grazing. The same applies to yard dogs and wandering stray dogs in rural areas. Our investigations of such animals from active agricultural foci showed a high degree of seropositivity against C. burnetii—49.45% (Chapter Eight).

12.8 Activity of Agricultural Foci and Epidemics of Q Fever in Humans

Permanent element in our works has been studying the relationship between the activity of the AF and outbreaks of Q fever in humans.

In Table 12.4, it is shown that during the last two epidemics of Q fever in humans, Botevgrad, 2004 and Etropole, 2002 compared retrospectively with the largest epidemic in the region of Knezha (1984) into the corresponding AF reservoir and source of infection to humans is clearly differentiated seropositivity for C. burnetii in three areas: *central* (focus of epizootic and epidemic); *neighboring* (first territorial belt without diseases in humans); and *distant*—second territorial zone, also without human cases. Serologically positive animals in the central zone are respectively 2.6 times (Botevgrad), 2.3 times (Etrophole), and 3.3 times (Knezha) more compared to the other, more remote areas of the country. These data not only demonstrate the relationship between epidemics Q fever in humans and corresponding AF of the disease, but can serve as a model for predicting possible outbreaks of Q fever among people when analyzing the sero-epizootiological records of the animals [620].

12.9 Relationships between Natural and Agricultural Foci of Q Fever

Between the two types of foci Q fever exists relationships and interconnections in several directions (Figure 12.6).

- The natural foci supported by several species ticks capable of transmitting the C. burnetii trans-phase and trans-ovarian from wild mammals and birds are vast reservoir with autonomous circulation of the infectious

Table 12.4 Comparative data for the serologic status of the infection with C. burnetii in foci of epizootics during epidemics of Q fever in humans

Seropositivity for C. burnetii in Domestic Ruminants (Type; Rates for Species; General Average %)

Foci of Epizootics in Ruminants and of Epidemics in Humans		First Territorial Belt (Areas Immediately Adjacent without Epidemics)	Second Territorial Belt (Remote Areas without Epidemics)
Area of Knezha (Vratsadistrict), 1984, 725 sick people			
Cattle	30.75%	15.03%	8.92%
Sheep	32.01%	14.35%	9.64%
Goats	23.11%	11.92%	8.76%
Total	30.97%	13.77%	9.11%
Etropole (Sofia district), 2002, 123 sick people			
Cattle	11.62%	8.79%	8.09%
Sheep	59.45%	20.86%	14.10%
Goats	100%	26.44%	13.81%
Total	25.42%	15.89%	10.91%
Area of Botevgrad (Sofia district), 2004, 220 sick people			
Cattle	33%	16.43%	8.41%
Sheep	46.66%	18.29%	9.90%
Goats	63.63%	23.48%	13.48%
Total	49.85%	18.86%	9.43%

agent. The latter is determined by the possibilities of cross-infection of their hosts that are implemented in the following ways:

a. By transmissible inoculation mechanism in bloodsucking of the ticks parasitic on the wild animals;

b. Per oral and inhalation uptake of the agent, excreted with the feces of ticks;

c. Alimentary and respiratory infection to susceptible wild mammalian and avian species with infectious secretions and excretions of already infected wild species.

- The agricultural foci of Q fever with their large and numerous reservoirs and vectors of infection among domestic animals also have their own circulation loop of the agent and two main ways of infection—respiratory and alimentary.
- Ticks in parasitizing on livestock in the external environment (pastures and roads) infect them by inoculation way through saliva during

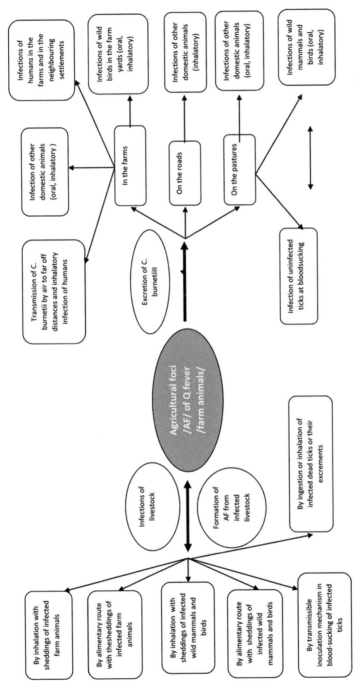

Figure 12.6 Circulation round of Coxiella burnetii in the agricultural foci and relationships with the natural foci of Q fever.

prolonged sucking. The arthropods transmit C. burnetii through their infected feces in their entry into the digestive tract and respiratory system of livestock at grazing.

- Similarly, infectious secretions of wild animals and birds serve as a source of infection for farm animals which are infected orally or by inhalation at pastures. In cases where carriers and emitters of the pathogen are wild birds, often seen in the courtyards of the farms where they find food (pigeons, doves, crows, magpies) is likely infecting with them to domestic animals in farms themselves. If these birds were not infected with C. burnetii but coexist with infected domestic animals, then it is possible the reverse process—contamination of the wild birds within the agricultural outbreak in taking contaminated with the causative organism feed.

- Affected by Q fever livestock emit C. burnetii in both farms and roads and pastures. They represent a reservoir, source, and vector of infection for susceptible wild mammals, birds, and ectoparasites in the environment. In some cases, when referring to free from the agent territory in nature, domestic animals appear source of infection for this area where getting infected with wild species, including ticks. Perhaps this is one way for the occurrence of new natural focus of Q fever. In other cases, domestic animals emit C. burnetii in already existing natural foci. This obviously leads to a refreshing of the natural foci with new quantities coxiellas and helps maintain and persistence of the infection in them.

    Apparently, in some geographical areas there is a systematic two-way circulation of C. burnetii between natural and agricultural foci and vice versa.

- The preservation of the agent of Q fever in natural and agricultural reservoirs, its circulation in both types of foci and bypass between them, is favored by a poly the adaptability of C. burnetii, their great resistance in the external environment, and the preservation of their virulence by drying and other physical and chemical impacts.

- Our observations show that despite the independent circulation round of the agent in AF, the availability in the vicinity of natural foci influences the course and intensity of epizootic process through the emergence of new infections and reinfection and contributes to the wider spread of Q fever among susceptible non-immune populations of animals and people.

12.10 Epidemiological Particularities in Humans

The main precondition for the occurrence of Q fever in humans is the presence of agricultural outbreaks of the disease which above were thoroughly characterized. The infected large and small ruminants are the main reservoir and source of infection for humans [118, 157, 198, 241, 323, 343, 381, 495, 634]. For human Q fever, the natural foci are not essential, since people rarely come into direct contact with them. In our research work, we have only one case of direct infection per person from fallen on it by tick species Dermacentor marginatus and transmitting the infection by bloodsucking. As a result, a state of fever, unilateral regional lymphadenitis and expressed serological response against C. burnetii are developed. In the tick was proved the presence of the infectious agent. The disease was influenced by the use of tetracyclines [597].

As observed by us, cases of Q fever in persons professionally engaged in animal husbandry and veterinary medicine clearly stood out the relationship between the epizootiological and the epidemiological chain of the disease. Particular risk for such personnel are clinically ill sheep, goats, and cattle releasing profusely the pathogen with their secretions and excretions. The infectious agent in these cases is transmitted by inhalation, by alimentary, or through contacts in the injured skin. As an example of the relationship and the common etiology of diseases caused by C. burnetii in shepherds and the sheep in their flocks in which there are miscarriages, stillbirths, non-viable lambs and retained placentas shows Table 8.15 (Chapter 8). In another well-documented case is concerned the sick T. G.—veterinary practitioner with generalized Q fever, manifested by massive lymphadenopathy and epileptic-like seizures. The infection is acquired in daily work in focus of Q fever in cattle and sheep where there have been many opportunities for contamination by contact mechanism of penetration through small erosions of the skin on the hands and through inhalation of contaminated with the agent dust or aerosol from livestock premises feed [611]. Inhalation infection in stay in pens with goats affected by Q fever manifested by abortions and respiratory disease was found in a veterinary surgeon D. K O., who developed an acute febrile condition affecting the liver and joints. In the blood serum of the patient was found accumulation of specific antibodies with seroconversion [597]. To diseases with professional character also relate a laboratory infection acquired by a veterinarian N.S., when working with a large number of blood samples and other clinical and pathological material from Q fever epizootic disease in cattle, sheep and goats, and the associated outbreak in humans (Knezha (Vr)).

In this case, the possible mode of transmission is combined—inhalation of infectious aerosols and by contact at the handling of blood, placenta, strains, etc. As a result, systemic infection was induced with intermittent fever, interstitial pneumonia, and accumulation of specific antibodies with high titers [597].

Described case examples are part of the classical epidemiology of Q fever in which the risk factor—work associated with immediate direct contact of man with sick animals or their secretions and products is a mandatory element in the mechanisms of transmission. In particular vulnerable to infection are newly employed workers and specialists in holdings representing active foci of Q fever.

Epizootic among ruminants is revealed by Knezha (1982–1985), Pavlikem (1985) Panagyurishte (1993), Etropole (2002) Botevgrad (2004), etc., and the smaller outbreaks (1995–2000)—Blagoevgrad, Sopot, Troyan, Pleven and related epidemics in humans, clearly established the link between the intensity of epizootic diseases in animals and the diseases in humans. In these epidemics has been deleted, the risk factor, respectively. Professional nature of diseases such as clinical forms of course, mainly pneumonia, occurred predominantly in individuals with occupations related to animal husbandry and processing of animal food and raw materials [599, 607, 608].

In the two major epidemics (Knezha and Pavlikem) were registered 1269 diseases. Of these, 725 were in Knezha and three neighboring villages as in the peak phase (1984) of the epidemic continued three years; the morbidity reached 2084% 000, while it was 6.56% 000 for the whole country [39]. The sick working in an environment with a maximum risk—holdings were about 5%, and engaged in other agricultural activities were approximately 11% of those affected. Over 80% of infected persons had various other professions, but lived in a small town and villages and raising animals in their small family farms.

The epidemic broke out amid extensive epizootic disease in farmed ruminants grown both in existing cooperative farms and in private farms, where it formed muster flocks of grazing. The conducted serological tests covered 5715 animals: 1826 cattle, 2414 sheep, and 1475 goats from 42 villages in the district of Vratsa, including 2244 from four settlements with cases of Q fever in humans (Knezha, Brenitsa, Lazarovo, Enitsa) with a number of affected persons, respectively 614, 79, 17, and 15 [586]. The remaining 3471 animals were located in two conditionally designated territorial belts: the first including the neighboring villages of these four settlements /foci/ and second covering the following and other more remote villages. The main part of the animal blood samples were taken in the period from late February to early June. The results showed a significant overall seropositivity for three zones—central (outbreaks of epizootic and epidemic), adjacent (first territorial zone without disease in humans), and remote (second zone, also without human cases), namely—15.08%, including 20.67% (sheep), 11.44% (cattle), and 10.44% (goats). The serological status of Q fever in animals in the central zone contains 4 outbreaks of epizootic and epidemic characterized by double larger overall percentage of positive seroreagents—30.97%. The percentage distribution by species is 32.01% (sheep), 30.75% (cattle), and 23.11% (goats). The highest range of infection in animals found in both places with the highest number of human diseases: Knezha—cattle (35.33%), sheep (34.59%) and Brenitsa—cattle (28.78%), sheep (32.59%). The spread of Q fever in the goats was almost equally—respectively 23.11% and 23.07% [586, 597].

In both zones outside affected by epidemics in humans, serologically positive animals were found in 16 settlements, and their number decreased in parallel with distance from the central zone.

These results indicate that the generalized spread of infection with C. burnetii including in the numerous personal holdings and its high activity are the main factors for the occurrence of disease outbreaks of Q fever in humans acquired by respiratory, alimentary, and contact route. The majority of cases, mainly pneumonias and bronchopneumonia, were found in March, April, and May, and the age of the persons most affected was from 30 to 50 years [39, 597].

On peculiarities of the course, the major epidemics of Q fever in Pavlikeni (1985) and Panagyurishte (1993) resemble the epidemic in Knezha. More clearly, the tendency for deviations from the typical seasonality in the past associated with the birth period in domestic animals and to shift the peak incidence of winter–spring to spring and summer months is outlined. Diseases were registered throughout the year, but the number was greater in April, May, June, and July. The epidemic in Panagyurishte, which covered more than 1,000 people, had its peak in June, July and August. Upon epidemic outbreak in Troyan (1997), a particularly high percentage of the patients was marked in May.

Between 1993 and 2002, there were prevalence sporadic cases or small outbreaks (Blagoevgrad, Sopot, Kyustendil, Troyan, Pleven) [599, 608].

Sporadic incidence of Q fever among the population registers all years. It is directly linked to sporadic cases of the disease in domestic ruminants, particularly cows, births in which there is no seasonality, unlike sheep and goats. In this way, excretion from the genital apparatus of the infected cows with pathology of pregnancy, or during normal birth, as well as the mammary gland and the digestive canal can be constant in any season of the year. The infected cows with reproductive disorders and continuous reduction in milk secretion also serve as a systematic source of infection for humans. Such role playing and dogs with Q fever—both in agricultural foci and wandering in the cities (Table 8.12). The infected cats pose a risk to the public and can be predicted strengthening their epidemiological importance in the coming years. Others, though minor so far, sources of infection for humans are animals with a dominant sporadic occurrence of Q fever—pigs, horses, and some poultry. Sporadic cases or infecting group of persons without direct contact with animals are possible through infected dust from hay, straw, grain, or feed containing dead ticks, rodents, or droppings of them after falling into the airways or the digestive tract.

The number of sporadic cases of Q fever among inhabitants of the cities, not having any contact with sick animals or foci, is convincing testimony to the changed epidemiology of Q fever. In these cases, obviously infections are caused by the agent of Q fever, spread to more distant distances with infected birds (pigeons, crows, migratory, etc.) by the so-called secondary sources of infection—contaminated with the pathogen feed, straw, and raw materials (skins, wool, bristles) or by winds carrying infectious aerosols and dust [597].

In 2002–2004 in Bulgaria marked a new upsurge in the epidemic incidence of Q fever outbreak by two epidemics in two neighboring region of Sofia region—Etropole, 2002 and Blagoevgrad, 2004. The serological and epizootiological studies on Q fever among domestic ruminants in both areas showed high incidence of the goats, sheep, and cattle (see Chapter Eight). Epidemiological information for the period March–June 2002 indicates that in Etropole were registered 123 cases of Q fever in humans as the incidence of the settlement reached 1038.41% 000 against the background of zero incidence in 2001 throughout the Sofia region and morbidity 0.70% 000 for the country. In 104 patients, the diagnosis was confirmed serologically. Affected were mainly people in active working age—from 20 to 50 years with diverse professions. The prevalence of Q fever has been identified in various points in the city. Over 50% of patients had not been engaged in rearing livestock. This fact gives grounds to assume that the most likely mechanism of transmission of infection in most patients was due to inhalation of infectious aerosols and dust. In some cases, especially in people raising animals cannot be excluded infection acquired through alimentary route—with milk, dairy products, or contaminated foods with secondary stumbled infectious secretions and excretions from animals, as well as contact-household time, especially when handling aborted animals, etc. Main reservoir and source of infection for humans are the domestic ruminants which in Etropole are located in small private farms. Traditionally, for the area has been pastoral raising sheep, goats, and cows, where they are in close contact with ticks and other natural reservoirs. Infected grazing animals are already agricultural reservoir and source of respiratory infection for people in their daily passing through the city streets. Clinically, the majority of patients are affected by pneumonia occurring atypically.

The analysis of the epidemic in Botevgrad and its adjoining area in May 2004 shows that the age structure of suffering from the disease 220 people is similar to that in Etropole—mostly people from 25 to 50 years, constituting 67% of those affected. Most patients—168 residents of the town of Botevgrad and the rest 52 live in nearby villages. The predominant clinical form is atypical pneumonia. The main route of infection is the respiratory, due to the strong dusting of the roads crossed by animals, going and returning from grazing. The beginning of the epidemic in Botevgrad was preceded by prolonged strong winds that caused extremely high level of dust which covered not only the roads but also the entire town. We assume that this impact of the external environment has served as a "trigger" factor for the outbreak of the epidemic by massive disseminating C. burnetii separated from the animals by air. Obviously, the high concentration of the inhaled agent caused respiratory disease in most susceptible individuals with an wavy increase in the number of those affected in the short term.

Another epidemiological point in the epidemic of Q fever in Botevgrad and Etropole was the role of the goats as the main source of infection for humans which is determined by their high prevalence, respectively. In the decade preceding the two epidemics the goat population was greatly increased. The role of sheep and cattle in the epidemiological chain is also a significant issue at these epidemics, as evidenced by the serological data in both species (Tables 8.3 and 8.5).

The incidence of Q fever, clinically manifested as atypical pneumonias occurring, tracking systematically during the period 1993–2004. The serological examinations covered 19,560 patients as with the exception of 2001 (633 patients), and the annual number of respondents ranged from 1139 to 2788 (Tables 12.2 and 12.3). For the entire 12-year period, the total amount of serologically proven pneumonias was 15.78%. The highest percentage (28.83%) was found in 1993, followed by 2004 (24.22%) and 2002 (20.51%). These are years that were recorded epidemics of Q fever.

The maximum incidence of pneumonia in 1993 was established in August (40.5%) and June (38.57%) to July (31.53%). In 2002, the highest monthly incidence was in August (73.94%), and in May, June, and July—respectively 35.25%, 26.55%, and 26.98%. The trend for the largest number of pneumonia caused by C. burnetii during the summer season witnessed in 2004—July (30.47%).

In non-epidemic periods, the proportion of pneumonia induced by C. burnetii was moving between 9.21% (2003)—9.39% (1999) and 19.72% (1996).

Comparing the incidence of lung inflammations associated with Q fever in 2001–2004—17.99% with the previous two four-year periods—1997–2000 (13.39%) and 1993–1996 (17.65%), shows almost identical percentages in the 2001–2004 and 1993–1996 years—periods recorded two medium-sized (2002 and 2004) and a large epidemic (1993). During the four-year period of 1997–2000, there were marked small outbreaks, but most cases of the disease were sporadic. Last among the group or mass infections were constant element in the structure of morbidity in these other two four-year period.

Aggregated data from Tables 8.2 and 8.3 (Chapter 8) show that pneumonia associated with C. burnetii occurs year-round with variations in the percentages reflecting the amount of this morbidity. General rates for each month of the year in the period 1993–2000, in descending order, were as follows: 22.08% (May); 21.73% (in July); 19.03% (August); 18.17% (May); 14.73% (April); 13.36% (March); 12.78% (September); 10.96% (November); 10.72% (December); 10.53% (February); 9.51% (October); 9.27% (January). There were three levels of morbidity: *first*, the relatively highest (from 22.08% to 18.17%)—in June, July, and August to May; *second*, medium (from 14.73% to 12.78%)—April February, and August; and *third*, relatively low (from 10.96% to 9.17%)—in October, November, December, January, and February.

The summarized monthly data on pneumonia associated with C. burnetii for four years (2001–2004) showed that 2/3 of this period (January, February, March, June, September, October, November, and December), this morbidity rate was in the range of 10.09%—13.8%. Dominated mortality rates in the range of ~10% (January, February, March, November), followed by ~12% (September, October) and ~13.8% (June, December). In the remaining four months, the incidence was higher: 15.54%–19.78% (April and June) and 31.85%–38.36% (May and August).

Besides the lack of a pronounced seasonality of Q fever in humans and deletion of risk factor, our studies also significantly expand the age range of those affected.

12.11 Analysis of Epizootiological and Epidemiological Data

Ticks, rodents, wild ungulates, and wild birds are susceptible to infection with C. burnetii. Known are dozens of types of ticks, mainly ixodic where it is demonstrated spontaneous infection with this microorganism [22, 30, 85, 128, 318, 405, 465, 515]. The species composition and geographical distribution of ticks—carriers of the agent, show wide variations. Of interest is the comparison of the above date on the natural reservoirs of Q fever with previous studies of other Bulgarian authors [20, 21, 471]. Georgieva [20] and Georgieva and Kiossev [21] in the late seventies of the twentieth century in the study of a total of 1891 ixodic ticks by IFHT establish carriage of the agent in 21.41% with variation from 10.1% to 50% for individual capture. In another study (IFHT) on a limited number of ticks, published in 1999, 25 (86.20%) out of 29 ticks had been infected with C. burnetii [471]. Highest degree of infection showed ticks collected from cattle (20.64%), followed very close by dogs (19.51%) and sheep (18.14%), while goats found 13.74 percent. By kind ticks, the grading rates of infection is led by I. ricinus (46.36%), and in descending order with small differences (27.65–21.15%) were R. sanguineus, R, bursa, B. calcaratus, D. marginatus, and H. plumbeum. Lower rates (from 11.11 to 9.74%) showed H. punctata and D. pictus. The data presented indicate that there is in Bulgaria a massive tick natural focal reservoir of Q fever with unquestionable role in the long-term and sustainable maintenance of outbreaks and circulation of C. burnetii in the external environment as well as systematic infection of wild and domestic animals [597, 606, 627]. Similar conclusions are made by other researchers [177, 318, 465, 491, 522, 525].

Very substantial are the results from the studies of natural focal animal reservoir of C. burnetii. Fournier et al. [198] indicate that the reservoirs of Q fever are wide, but partially known, and include mammals, birds, and arthropods, especially ticks. The authors define the small wild rodents as an important natural reservoir of the infection [198]. We share the opinion of the researchers cited for the insufficient knowledge on possible other reservoirs infection with C. burnetii in the animal world. This undoubtedly requires the need for continued research quest in this direction. Studies focused on 17 species of wild animals and birds from different regions, most of which have not been subject to etiologic research in the country [589, 597, 628]. Conducted serological tests of nearly one thousand blood samples revealed the presence of specific antibodies against C. burnetii in 15 species: hares, foxes, three species of ruminants, six species of small rodents, and four bird species. The registered high rates of seropositive reactions to the role of wild rodents maintains infection in nature. With obtained serological evidence of infection with C. barnetii in deer, hares, ground squirrels, crocidura (Sorex), and magpies confirm results of other Bulgarian authors from the late fifties [65, 66], sixty [16, 40, 41], and early seventies [14, 17] of the last century. The serological results were convincingly supported by direct detection of C. burnetii and isolation of 12 strains of the agent from seven types: forest dormouse (first description in the literature); pigeons, doves, crows, and pheasants [597, 628] and wild rabbit—confirming previous results [14, 17]. Shindarov et al. [66] and Serbezov [57] reported for isolation of C. burnetii from ground squirrels; Genchev et al. [17] and Ganchev et al. [14]—from wild rabbits; and Genchev et al. [17]—from badger. The infectious causative organism reproduces in infected wild species and is excreted in their secretions and excretions. Thus, they serve as sources of infection for other animals on the field and ticks. Thus, apart from the numerous types ixodic ticks, a number of wild mammals and birds maintain extensive active foci of infection with C. burnctii with high virulence and toward domestic species of animals, birds, and people in contact with them [46, 85]. Stein et Raoult [495] reported an outbreak of acute Q fever (pneumonia) affected family of five living on the farm in Provence, France. The epidemiological study revealed that the outbreak was the result of contact with contaminated with coxiellas feces of pigeons regularly overflying and landing at the farm family [495]. This example shows that the bird natural reservoir of C. burnetii under certain conditions can serve as a carrier of direct infection of humans.

The issues of the farm outbreaks and foci of the infection and related epidemiological characteristics of Q fever are essential for deep insight into the nature of this zoonosis and defining the strategy to combat it. The systemic and large-volume serological and virological tests have enabled us to establish many foci of Q fever among domestic ruminants with different spatial locations. The analysis of the obtained data shows that the continued reveling of new foci of infection is related to cattle, sheep, and goats. No less important moment in the studies of the problem is monitoring of status, activity, and dynamics of previously known foci of the disease.

The overall view on nosological geography of Q fever in Bulgaria stands a characteristic of the different regions, namely that there are many villages and farms that are free from infection with C. burnetii. This fact is essential in organizing the prevention and fight against Q fever, emphasizing the importance of mandatory preventive serological testing of flocks groups and individual animals before moving them to other farms as well as acquisition of herds within a farm.

We believe that the detailed characterization of agricultural foci Q fever is a complex task requiring a multilateral approach and complex analyses and evaluations. In this respect, differentiated relevant indicators and actions we implemented in several directions: (a) way of raising animals and organizational structure of livestock farming; (b) number, density, distribution, and localization of the agricultural foci on the territory of a settlement, district, or region; (c) presence of single foci of Q fever or conglomerates from them; (d) internal structure of the foci depending on the type of herds—standing with based in separate buildings for breeding large number of animals, or pooled with daily collection and return to private yards; (e) categorization of animals in foci, depending on their clinical status and excretion of the pathogen; (f) level of activity in the outbreaks of Q fever.

Obviously, the proportions of those categories and elements of the agricultural foci are diverse and dynamic, which excludes one-size schematic approach in the epidemiological studies. We define this type of foci as complex changing multicomponent systems with a range of features and relationships. We believe that a critical characteristic of the once created agricultural foci is to maintain a separate conduit of C. burnetii and cyclical reproduction of the infection. Similar opinion was expressed by Rehaček et al. [431] based on surveys of Q fever in cattle in Germany. As people are much rarer contacts with the natural foci of the disease, the agricultural reservoir of the pathogen plays a key role in the epidemiology of Q fever in humans [46, 69, 167, 198, 226, 305, 353, 509, 564].

The question of persistence of C. burnetii in the organism of the animals and their excretion with their secretions and excretions is of great importance. It is generally recognized for very strong discharge and disseminating infection in the maternity period and in the case of abortion and related conditions [44, 85, 277, 456, 509, 525]. No less risk regarding the release of the agent are the other clinical forms of Q fever in animals, particularly the mastitis and reproductive disorders. It is considered that mammary gland and uterus are the main localizations of the pathogen in a chronic course of the disease. Several teams of authors found that the cows more often than the sheep develop chronic infection with C. burnetii and longer emit the agent with milk, which defines them as the most important source of infection for humans [85, 111, 277, 288, 343]. Paiba et al. [385] indicate that C. burnetii infection is of great importance for public health in England and Wales, and it is related to dairy cows. The authors reported that 21% of milk samples from randomly selected farms showed serological evidence of infection with this organism. Kim et al. 2005 [277] recently published data from the 3-year study by PCR of milk samples from cows in the USA, indicating that over 90% of herds for milk were infected with the agent of Q fever. The extrapolation of the data over national dairy herd in the US means that approximately 3 million lactating cattle daily release C. burnetii in milk [277].

The goats, like the cows reared for milk, indicate a predisposition to the development of chronic infection with C. burnetii [288]. The significant increase in their number as a source of milk at the expense of the smaller number of cows poses a risk to people of transmission of the infection [194, 288, 414, 447, 471]. Bulgaria is one of the examples in this respect over the past decade by nearly threefold increase in the number of goats raised in numerous fragmented private farms. Continuing their passage through the streets of the villages on the way to pastures and back leads to a severely dusting. These animals are in daily close contact with the pooled flocks of goats and with the people in homes [594]. That practice of growing concerns and sheep on the small personal farms, where their number is usually from 2 to 10 animals. The results of the above-cited studies [81, 198, 305, 336, 468, 530, 564] have shown that the sheep are massive agricultural reservoir of C. burnetii of considerable economic and epidemiological importance. In the studies on the epidemiology of Q fever must also taken into consideration other particulars: changes in the semen practices: seasonality of maternity periods—in sheep (December/February) in goats (mainly in March/April) and the cows (year-round); deficiencies related to uncontrolled movement of animals from infected to disease-free zones and vice versa; the broad possibilities for respiratory and alimentary infection of the animals and so on.

Besides cattle, sheep, and goats, our serological tests for Q fever covered a number of other domestic animals and birds—a total of 8169 of 9 species [597]. For such research in buffaloes, horses, pigs, camels, rabbits, dogs, cats, and poultry were reported in other publications [242, 252, 269, 338, 522], which reveal varying degrees of receptivity and scope of infection. According to American authors [525], each species can be infected with C. burnetii under natural conditions, but the affected cattle, sheep, and goats are the main risk factor for humans. We revealed relatively poor circulation of the infectious agent among buffalo population in the country (prevalence of 1.43% in the three areas) and the lack of positive seroreagent in other areas. For those results they contribute likely and natural resistance of buffaloes to infection with C. burnetii. When conducted systematic studies of horses and pigs rarely discovered seropositive animals and the total prevalence in both species was negligible (<1%). Based on these results currently participation of horses and pigs in the formation of agricultural foci Q fever in the country should be excluded, but it is necessary for continuation of sero-epidemiological surveillance, especially in areas where there are active foci of infection in other species. The positive findings in poultry suggest the need to extend the objects of sero-epidemiological surveillance of Q fever with a view to the timely disclosure of additional reservoirs and sources of infection.

For dogs have long been assumed that participate in the formation of agricultural foci of the disease [314]. These animals are in close contact with both holdings and the man and his home. The high percentage of infected sheepdogs and yard dogs with C. burnetii in our survey clearly shows that they alone or by ticks that carry serve as an important reservoir and source of infection for other animals and humans. In this respect, our views coincide with the opinions of Boni et al. [103], Buhariwalla et al. [115], Laughlin et al. [290], and Rauch et al. [429]. The problem of the large number of homeless dogs in the cities is a signal of serious threat of infection of humans by direct or indirect contact with these animals. Similar is the situation with the stray cats [278, 606]. Susceptibility of cats to C. burnetii is underscored by the discovery of infected animals reared in rural and urban homes. Contact with such cats, especially in the birth period, can result in diseases of the humans [198, 242, 289, 319, 338, 399]. Excretion of coxiellas resistant to drying and into the urine, feces, and milk is also a risk factor for transmission of infection from cats to humans [198]. The facts attest to the undeniable zoonotic importance of cats and dogs that in terms of the urban environment are an important reservoir of C. burnetii and source of infection in urban outbreaks of Q fever among the population. In our opinion, in the case of detection of C. burnetii infection of dogs and cats in large cities—as ownerless and raised at home, it is correct using the term *"urban foci of Q fever"* [597].

For a more complete understanding of the problem Q fever and more effective countermeasures crucial is the need for knowledge and deep insight into the complex relationships and mutually conditioned links between the two main components of epidemiological point of view—the existence of natural and agricultural foci of disease [597, 632].

The epidemiology of Q fever in humans reflects the epizootiological characteristics of the disease and the biological nature of the etiological agent. The reservoirs and sources of C. burnetii infection are diverse, and the clinical conditions induced in susceptible individuals vary widely—from headaches to death [545]. In connection with the classical epidemiology of Q fever in humans were identified key risk occupational groups—workers in natural reserves [509]; livestock and veterinary staff [28, 118, 167, 169, 224, 414, 517, 530, 564]; workers in industries processing food products and raw materials of animal origin and animal health inspectors in them [118, 380, 408, 414, 527]; staff research and diagnostic laboratories in the field of microbiology, virology and immunology, and experimental bases with animals [212, 228, 352, 394, 509, 527]; troops in terms of teachings and actions on the field [3, 118, 127, 190, 213, 292, 310, 483, 541]. Very important in contemporary conditions are observed deviations from traditional epidemiology of Q fever related to seasonality, risk, age of affected individuals, and others [241, 426]. Hellebrand et al. [241] carried out extensive epidemiological studies in Germany indicate that seasonal outbreaks of Q fever among the population has changed from predominantly winter–spring in predominantly spring–summer, most likely due to changes in sheep. Localization of recent epidemics in this country leads to the assumption that the urbanization of rural areas contributes to the rise of the incidence of Q fever [241]. The radical changes in agriculture in each country relating to ownership, the organizational structure of the industry and manufacturing practices in the cultivation, and breeding of animals affected the distribution, seasonality, and other epidemiological characteristics, and this is reflected in the epidemiology and clinical course of the Q fever in humans.

Raoult et al. [427] indicate that in the majority of publications on the epidemiology of Q fever, inhaled route of infection is considered a major for humans. Without a doubt, the proven high resistance of C. burnetii in the external environment is favorable factor for long-term survival of the pathogen and its spread from the outbreaks of infection. Lyytikänen et al. [305] indicated that the formation of infected with the agent aerosols and dusts is facilitated in dry weather. The air currents—natural winds or created by human activities (landings of helicopters, passing trucks transporting animals and feed, demolition of old buildings, etc.) carry the infectious agent from the agricultural foci or infected slaughterhouses and other objects at different distances where you be realized respiratory infection [81, 118, 169, 224, 299, 517, 530, 533, 564]. Role in the transmission of the infectious cause of greater distances, including the cities, probably play some birds—crows, magpies, pigeons, migratory, and others. Interesting data on the outbreak of Q fever in industrial enterprise (unrelated to livestock farming or processing of products) was presented by Van Woerden et al. [533]. The authors found that the source of infection are thatched ceilings in the old building at the destruction of any form dense clouds of dust with subsequent infections of humans. The straw was contaminated with coxiellas from the carcasses of infected wild brown rat (Rattus norvegicus) and their feces, urine, or birth products [533]. For these rodents is known to be an important source of C. burnetii infection for farms in the UK, which leads to zoonotic risk [546, 547]. According to Comer et al. [141], ordinary rodents in urban areas can also contaminate the straw through their placentas contained large amounts spores—like forms of C. burnetii. Alimentary route of infection with C. burnetii is also important, especially for people raising animals and their families after eating infected raw cow, sheep, and goat milk and goat cheese [93, 194, 285, 355, 509]. Tissot Dupont et al. [517] indicate that the rural population greatly decreased in recent decades, so now Q fever is common among urban residents, especially after casual contact with infected animals or from eating contaminated with C. burnetii raw milk. In our opinion, it is possible alimentary contamination with other foods contaminated from the secondary stumbled secretions and excretions of animals affected by Q fever. In this regard, the assessment of risk of infection of people with C. burnetii in consumption of potentially risky foods is essential [625].

In connection with the age of the people affected by Q fever have different opinions [343, 355]. Fournier et al. [198] emphasize that Q fever affects all ages. The actual incidence and the age range appear to vary, because of the undoubted passing the majority of cases, since diverse and non-specific clinical signs are not always directed to the disease [118, 198, 343]. The omission of many cases of Q fever is particularly common in children [273, 313, 449, 521]. The omission of many cases of Q fever is particularly common in children [273, 313, 449, 521]. For example, studies in Japan [521] establish the importance of C. burnetii as an etiological factor in pneumonia in children (39.7%). These data and other information not mentioned once again confirmed the change in the epidemiology of Q fever and necessitating regular etiological diagnosis of the disease with coverage of a wider range of age groups.

13

Sensitivity to Antibiotics and Treatment

In the complex of measures to control and fight against Q fever in animals and humans, treatment and drug prevention occupy an important place. The search began in the 1940s of the twentieth century, shortly after the proof of the disease and identification of the etiologic agent. Inability to cultivation of C. burnetii in artificial culture media makes it impracticable for the conventional analysis to determine the sensitivity to antibiotics. In this regard, three fundamental experimental systems for studying this question are designed: chicken embryos (in vivo), animal (in vivo), and cell culture (in vitro).

The activity of antibiotics against C. burnetii is judged by their ability to prolong the mean survival time (MST) of infected CE at inside yolk inoculation, making comparisons with control CE not treated with antibiotics [343].

Another indicator of the therapeutic action of the antibiotics to chick embryos is reduced presence or absence of the causative agent, established after the passage of the suspensions of YS of the survivors CE in uninfected embryos. According to Huebner et al. (1948), the treatment of CE infected with strain Nine Mile, with high doses of streptomycin—10 mg per egg—leads to an extension of the MST of six days, but C. burnetii is not eliminated from the embryos. Smadel et al. (1947) and Wong et al. (1950) found the curative effect of aureomycin and tetracycline on coxiellas in experimental infections of CE. In another early study (Jackson, 1951) has been reported a positive effect of the application of aureomycin, terramycin (oxytetracycline), and chloramphenicol.

Testing of antibiotic activity against C. burnetii of a model CE is possible through the application of CFT on the basis of registered slowdown of the positive complement fixing values and by comparing the CF titers of treated and untreated embryos with antibiotics (Ormsbee and Pickens, 1951). The results of this study show a good effect of oxytetracycline, relatively good

239

effect of aureomycin, unsatisfactory effect of chloramphenicol at a dose of 1 mg/egg and streptomycin (5 mg/egg), and the absence of inhibitory activity of penicillin G (6 mg/egg).

In the later period—the early eighties—Spicer et al. [484] tested the effectiveness of some antibiotic suppressions of mortality in CE experimentally infected with C. burnetii, Rickettsia rickettsii, and Rickettsia typhi. The authors found that the surveyed American, Scottish, and Cypriot strains are sensitive to the action of rifampin, trimethoprim, doxycycline, and oxytetracycline, but subsequent cultivation in YS of CE shows only inhibitory but not eliminating effect [484]. The quoted strains had been resistant to erythromycin, biomycin, clindamycin, cycloserine, and cephalothin [484]. The same publication reported for the first time on a natural resistance to isolate "Cyprus" of C. burnetii to tetracycline [484]. Raoult et al. [420] held in vivo experiments with fluoroquinolones (ofloxacin and pefloxacin) and found to be effective against strain Nine Mile of C. burnetii.

Experiments of animal experimental systems suggest that guinea pigs are very suitable for the purpose of the studies. The first attempts date back to 1948 when Huebner et al. treated with streptomycin the guinea pigs that are infected with the strains Henzerling or Dyer. There has been a positive impact, but only at high doses—40–50 mg/kg bw. In the same year (1948), Wong et al. reported that aureomycin is effective for the treatment of guinea pigs that are experimentally infected with C. burnetii. The described experimental models (mice and guinea pigs) to study endocarditis caused by C. burnetii appear to be promising for the identification and susceptibility to antibiotics, but still lacks certain data in this regard [287, 343].

The activity of antibiotics against intracellular arranged coxiellas was evaluated in vitro in various cell lines: murine fibroblasts (L929); murine macrophage-like cells (J774; P388D1); Vero cells; Hell(HeLa) cells, and others. A significant part of the initial research on this model was made in the 1980s of the twentieth century. In most cases, the antibiotic activity is measured by determining the percentage of infected cells using light microscopy and the IF in antibiotic-containing cultures and compared to control cultures without the presence of antibiotic [420]. Martin et al. [341] proposed a quantitative model to calculate a dose of the infectious inoculum cultures before and after the addition of antibiotic. Brennan and Samuel [107] determine the susceptibility of C. burnetii towards antibiotics by analysis, referred to "Real-Time PCR Assay."

Antibiotics of the tetracycline order show a pronounced effect against C. burnetii in most studies conducted in vitro in cell culture systems. Chopra and Roberts [137] presented a comprehensive overview of the tetracyclines—mechanism of action, molecular biology, application and the resistance of Coxiella burnetii. Along with this continued the research quests on the sensitivity of C. burnetii to antibiotics of other groups. Gikas et al. [206] examined the Vero cell efficiency of linezolid against Greek isolates of C. burnetii from patients with acute Q fever, along with similar research on quinolones, doxycycline, and clarithromycin. The mean inhibitory concentrations (MICs) of linezolid and clarithromycin ranged from 2 to 4 µg/ml of doxycycline, ofloxacin, and trovaxacin from 1 to 2 µg/ml, pefloxacin from 1 to 4µg/ml, and ciprofloxacin from 4 to 8 µg/ml. It is concluded that the linezolid is effective against intracellular C. burnetii in the cell line used [206]. Another team of authors—Rolain et al. [443] studied the bacteriostatic and bactericidal activity of moxifloxacin against C. burnetii. Maurin and Raoult [342] found in the model that HeLa cells are more sensitive isolates of C. burnetii, causing acute Q fever (including Nine Mile) and chronic form of the disease (Q212, Priscilla, etc.) to the fluoroquinolone levofloxacin than to ofloxacin. In experiments on lines L_{929} and P_{338} D_1 chronically infected with strain Nine Mile, doxycycline (4 µg/ml), rifampin (4 µg/ml), and pefloxacin (1 µg/ml) have shown a bacteriostatic effect and prevented cell death due to the inhibition and reproduction of C. burnetii, and lack of bactericidal effect, and the percentage of the infected cells remains unchanged. Using a combination of doxycycline and agents exhibiting tropism to lysosomes, leading to alkalization of the cellular environment—chloroquine or amantadine results in a synergy and elimination of C. burnetii of the cells until the ninth day [417].

The bacteriostatic action of antibiotics and chemotherapeutics is based on the inhibition of the protein synthesis of microorganisms by a variety of mechanisms [4]. Under the influence of doxicycline, the cell walls of C. burnetii become damaged. The fluoroquinolones inhibit microbial DNA gyrase involved in the mechanism of information processes of the DNA of this microorganism. The trimethoprim and combined preparation co-trimoxasolum inhibit and block the microbial metabolism of the folic acid [4].

Important is the question of the resistance of C. burnetii to antibiotics. In the early nineties have been found variations in the response of different isolates in respect of doxycycline, ciprofloxacin, rifampicin, and erythromycin [4, 420, 571]. The mechanisms of resistance of C. burnetii to antibiotics are not decrypted. It is assumed that a role in this phenomenon plays the permeability of cells of this infectious agent [89] and particularly in the quinolones—the nucleotide mutations [370]. Perhaps the use of different strains and isolates of the pathogen than previously tested will show marked heterogeneity in susceptibility to various antibiotics and varying degrees of resistance.

Very significant is the correlation between the intracellular sensitivity of C. burnetii to antibiotics at a level experimental systems and the effectiveness of their clinical application. Although *in vitro* experiments have significant activity of the tetracycline group, trimethoprim, rifampicin, fluoroquinolones, clarithromycin, and other newer macrolides, clinical data show good agreement with a part of them—tetracycline, quinolones, and trimethoprim [343]. In this respect, there are many other examples of incomplete or missing correlation of results in both of those criteria [118, 343, 393, 479, 524].

To treat acute Q fever in humans, doxycycline is the most appropriate antibiotic preparation—"drug of choice" [3, 4, 126, 129, 264, 353, 378]. Tetracycline is also often used [4, 118, 126, 129, 355].

The successful *treatment of chronic Q fever (endocarditis)* is much more difficult, continuous, and requires a combination of drugs: doxycycline with quinolones for not more than a short period of 4 years [425]; doxycycline with hydroxychloroquine for 1.5–3 years [126, 309]; tetracycline or doxycycline plus trimetoprim-sulfamethoxasole or rifampin for 12 months or longer [129]. Often, however, is required the replacement of affected heart valve [126, 129]. Summarized data for the advanced treatment of endocarditis caused by C. burnetii are published by Eliot et al. [180].

It applies also urgent prophylactics (preventive treatment) of Q fever [3, 4, 129]. Alexandrov et al. [3] developed a bicyclic scheme based on doxycycline for preventive treatment of Q fever and rickettsioses—tick-borne-spotted fevers, epidemic and endemic typhus.

The information on the use of erythromycin is contradictory. Some authors [393, 479, 524] reported good effect in the treatment of atypical pneumonia. In other observations, there was no similar effect [324]. Overall, it has the opinion that erythromycin should not be considered a reliable alternative in the treatment of Q fever [343]. Rather, this antibiotic has the potential for use in pregnant women and children [343]. Data on clinical experiments with newer macrolides (roxithromycin, clarithromycin, azithromycin) are scarce [118].

The information set forth attests to the complexity of the problem of the sensitivity of C. burnetii and antibiotic treatment of Q fever. There is a clear need to continue research efforts in this direction by covering new aspects of research—the role of the phase state of strains C. Burnetii, deepening the molecular biological studies and clinical trials of antibiotics in infected with different isolates experimental and domestic animals.

Contemporary data of OIE-CFSPH [380] suggest that little is known about the effectiveness of the antibiotics in the treatment of Q fever in ruminants and other domestic animals. Antibiotic therapy is recommended to reduce the risk of abortion and excretion of the pathogenin, the external environment [380]. Research is needed on the bacteriostatic and bactericidal action of the antibiotics against C. burnetii in their clinical application in veterinary medicine, selection and testing regimens, and urgent prophylactics (preventive therapy) and to assess the overall epidemiological effect of using them [122, 156, 380]. The literature review shows that the majority of developments in the application of antibiotics and chemotherapeutics are the work of human medicines researchers. It is clear that new research efforts to design and test schemes for treatment and antibiotic prophylaxis and their clinical application in veterinary medicine are necessary.

In present monograph, we made an effort to meet the needs of modern research on the issue of drug treatment of Q fever in animals. An important methodological point was the selection of therapeutic regimens according to the defined goals of treatment with antibiotics. The results show that the overall assessment of the application of tetracyclines in sheep, goats, and cattle is positive with respect to limiting the scope of the clinical disease and often for their termination. The same goes for the obvious preventing or limiting the development of the clinical forms showing original trend for mass scale in the active foci of infection. The third positive effect refers to the reduction in excretion of C. burnetii with genital extrudes and other secretions and excretions in abortion and related conditions, endometritis, metritis, mastitis, pneumonia, etc., which leads to a lower degree of contamination of the external environment.

We find suitable and convenient for using tetracycline preparations with prolonged duration of action (24 hours, 3 days), injected intramuscularly. The easier mode of application provides opportunities for the coverage of larger groups of cattle and whole herds of small ruminants when there are indications of their treatment. Too indicative are the results of the oral administration of chlortetracycline in feed premix during the dry period in spontaneously infected cows. Month-long acceptance of the antibiotic eliminates the agent in the genital secretions, urine, milk, and blood. A very good effect of oral treatment of chronically infected dry cows with chlortetracycline was announced by Behymer et al. [92]. The authors found no C. burnetii in the placenta and colostrum after birth or in the milk of treated animals [92]. Note that with tetracycline, antibiotics were not only used to treat clinically ill sheep, goats, and cattle, but also to treat the close contact endangered animals (preventive therapy). The prevention of clinical signs in these animals is definitely an evidence of successfully implemented antibiotic prophylaxis of Q fever, similar to that described for the man [4]. Studies have shown that the elimination of carriage and excretion of the agent needed prolonged and repeated treatment courses, and control laboratory tests. It must be emphasized that antibiotic therapy and urgent prevention are undoubtedly needed by epizootiological, zoo-prophylactic, and economic considerations. Too revealing in this regard is our experience with epizootic diseases of Q fever in domestic animals and related epidemics in humans during the period 1984–2011.

Despite the relatively good effect on the clinical use of tetracycline antibiotics, scientists continue to carry out studies on the effect of alternative antibiotics and chemotherapeutics. The encouraging data from the laboratory tests of some fluoroquinolones were followed by our first attempts for the clinical application of pefloxacin and ciprofloxacin in infected flocks of sheep with mass abortions and births of puny unviable lambs. Received the first positive results attest to the prospects of continued clinical trials and studies on the effect of the preparations on the carriage and excretion of C. burnetii from the animals. The data presented attest to the complexity of the problem of the sensitivity of C. burnetii and the antibiotic treatment of Q fever. There is a clear need to continue research efforts in this direction by covering new aspects of research—the role of the phase state of the strains C. burnetii, deepening of molecular biological studies and clinical trials of antibiotics and chemotherapeutics in experimental and domestic animals infected with various isolates. Part of our scientific research on one of these issues is presented in detail in the next section.

13.1 Sensitivity of C. burnetii to Certain Antibiotics Depending on Its Phase State

This question is explored experimentally. In our experiments, we used C. burnetii in a different phase state: Phase I—strain HP (miscarriage in cattle), a transitional phase—strain MP (isolation from placenta of a cow with a normal birth); Phase II—strain PM (mastitis in goats) [631].

The sensitivity of these strains was tested using two quinolone preparations—pefloxacin (1-ethyl-6-fluoro-1, 4-dihydro-7-(4 methyl-piperazine-1 IL-4-oxo-3-quinoline carboxylic acid) and ciprofloxacin (1-cyclopropyl-6-fluoro-1, 4-dihydro-4-oxo-7-piperazine-1-IL-quinoline-3-carboxylic acid), to the macrolide rulid (roxythromycin) and to the aminoglycoside amikacin (amikacin sulfuricum) [629].

Antibiotic activity of each preparation is tested in four concentrations, 0.001, 0.0005, 0.00025, and 0.00012 g/ml aqueous solution, and for each experiment inoculated into the yolk sac by 20 CE at 7 days of age at a dose of 0.5 ml 10% suspension of YS infected with the corresponding strain. Twenty-four hours later, the embryos were inoculated with 0.5 ml aqueous solution from the corresponding concentration of the test antibiotic. On the ninth and twelfth day, the infection kills two of the remaining living embryos from each test group, extracting the yolk sac and their contents diluted with saline in the ratio 1: 1, with the so-obtained suspension intraperitoneally inoculated at a dose of 4 ml of 2 guinea pigs. Temperature was measured for the test animals for 30 days, after which they were examined serologically.

The results obtained are shown in Tables 13.1–13.3.

Table 13.1 related to C. burnetii in the first phase shows that most are the surviving embryos after the treatment with 0.001 g and 0.0005 g ciprofloxacin—78.9%, respectively, 76.5%. The situation is similar in the test group treated with the same amount of pefloxacin—77.8%, respectively, 73.6%.

At the same dosage (0.001 g and 0.0005 g), roxithromycin prevents a small percentage of embryos, 59.5% and 29.4%, respectively.

In lower concentrations (0.00025 g and 0.00012 g) the above three antibiotics had little protective effect—from 5.5% to 38.9%. It is noteworthy that at a concentration of 0.00012 g, roxithromycin prevents a higher percentage of embryos (10.5%) compared to ciprofloxacin and pefloxacin (5.5%). Only amikacin does not prevent or slow down the dying of all embryos in their treatment with the investigational antibiotic concentrations (Table 13.1).

The part of Table 13.1 related to guinea pigs showed different titers of complement-fixing antibodies against C. burnetii, which indicates that inoculated suspensions of YS have had an uneven amount of the pathogen. Lowest serological titers (1:10) was established in guinea pigs inoculated with 0.001 g and 0.0005 g of ciprofloxacin and pefloxacin, respectively.

The disease in these animals proceeds inapparent. Conversely, in the guinea pigs treated with lowered concentrations of these antibiotics, as well as with all concentrations roxithromycin, we found higher CF titres—up to 1: 160. Their disease was accompanied by a rise in the internal body temperature for a period of one to three days, usually two days (Table 13.1).

The results in Table 13.2 on the sensitivity to antibiotics of the agent in transitional phase showed that most are living embryos treated with 0.001 g and 0.0005 g of pefloxacin, respectively, 89.4% and 84.2%. In the group of CE treated with the same concentrations of ciprofloxacin, these indicators were 84.2% and 82.3%, respectively.

Table 13.1 Sensitivity of *C. burnetii* strain HD (*first phase*) to pefloxacin, ciprofloxacin, roxithromycin, and amikacin

| Type of Antibiotic | Amount of Antibiotic | Chicken Embryos | | | | | | | | | | | Guinea Pigs | | | |
| | | Days and Daily Mortality after Infection with the Strain (No.) | | | | | | | | | Total No. of Dead CE | Live CE (No./%) | 9th Day | | 12th Day | |
		4	5	6	7	8	9	10	11	12			Days of Fever (No.)	CF-titer	Days of Fever (No.)	CF-titer
Pefloxacin	0.001	—	—	—	—	—	—	—	2	2	4	14 (77.8)	0	1:10	0	1:10
	0.0005	—	—	—	—	—	—	—	2	3	5	14 (73.6)	1	1:10	1	1:20
	0.00025	—	3	2	2	3	1	—	—	—	11	6 (35.3)	2	1:20	1	1:20
	0.00012	4	4	5	3	1	—	—	—	—	17	1 (5.5)	2	1:40	3	1:160
Ciprofloxacin	0.001	—	—	—	—	—	—	—	3	1	4	15 (78.9)	0	1:10	0	1:10
	0.0005	—	—	—	—	—	—	—	2	2	4	13 (76.5)	0	1:10	0	1:10
	0.00025	—	—	—	—	—	1	4	3	2	11	7 (38.9)	1	1:40	2	1:40
	0.00012	—	5	3	4	5	—	—	—	—	17	1 (5.5)	2	1:40	3	1:160
Roxithromycin	0.001	—	—	3	—	2	2	1	3	—	8	10 (55.5)	1	1:10	1	1:10
	0.0005	—	—	3	2	4	1	2	—	—	12	5 (29.4)	2	1:20	2	1:40
	0.00025	—	3	4	3	3	1	—	—	—	14	4 (22.2)	1	1:40	2	1:40
	0.00012	4	5	3	4	1	—	—	—	—	17	2 (10.5)	2	1:80	2	1:80
Amikacin	0.001	—	2	2	3	5	2	1	2	—	17	—	—	—	—	—
	0.0005	—	3	5	3	2	5	—	—	—	18	—	—	—	—	—
	0.00025	—	5	4	4	5	1	—	—	—	19	—	—	—	—	—
	0.00012	4	4	3	4	2	—	—	—	—	17	—	—	—	—	—
Controls																
Strain PM	10⁻⁴	5	4	5	5	5	—	—	—	—	19	0	5	1:640	—	—
Pefloxacin	0.001	1	—	—	—	—	—	—	—	—	1	18	—	—	—	—
Ciprofloxacin	0.001	1	—	—	—	—	—	—	—	—	1	18	—	—	—	—
Roxithromycin	0.001	—	—	—	—	—	—	—	—	—	—	19	—	—	—	—
Amikacin	0.001	—	1	—	—	—	—	—	—	—	1	19	—	—	—	—
Physiological saline	0.05 cm³	4	—	—	—	—	—	—	—	—	—	20	—	—	—	—

Table 13.2 Sensitivity of C. burnetii strain MP (transitional phase) to pefloxacin, ciprofloxacin, roxithromycin, and amikacin

Type of Antibiotic	Amount of Antibiotic	Chicken Embryos — Days and Daily Mortality after Infection with the Strain (No.)									Total No. of CE		Guinea Pigs — 9th Day		12th Day	
		4	5	6	7	8	9	10	11	12	Dead CE	Live CE (No./%)	Days of Fever (No.)	CF-titer	Days of Fever (No.)	CF-titer
Pefloxacin	0.001	—	—	—	—	—	—	1	1	—	2	17 (89.4)	0	1:10	0	1:10
	0.0005	—	—	—	—	—	—	—	2	1	3	16 (84.2)	0	1:10	0	1:10
	0.00025	—	—	—	—	2	2	1	3	—	8	10 (55.5)	1	1:10	2	1:20
	0.00012	—	—	4	5	3	1	—	—	—	13	4 (23.5)	2	1:20	2	1:40
Ciprofloxacin	0.001	—	—	—	—	—	—	1	1	1	3	16 (84.2)	0	1:10	0	1:10
	0.0005	—	—	—	—	—	—	—	1	2	3	14 (82.3)	0	1:10	0	1:10
	0.00025	—	—	—	—	—	3	2	4	—	9	10 (52.2)	2	1:40	2	1:20
	0.00012	—	—	4	3	3	4	1	1	—	16	3 (15.8)	1	1:40	2	1:80
Roxithromycin	0.001	—	—	—	—	—	—	2	2	1	5	13 (72.2)	0	1:10	0	1:20
	0.0005	—	—	—	2	3	4	2	2	—	11	8 (42.1)	1	1:10	2	1:20
	0.00025	—	—	—	2	3	3	3	—	—	11	6 (35.3)	1	1:20	2	1:20
	0.00012	—	—	4	3	3	4	—	—	—	14	2 (12.5)	1	1:40	2	1:80
Amikacin	0.001	—	2	4	3	1	4	3	—	—	17	2 (10.5)	—	—	—	—
	0.0005	3	4	6	3	3	—	—	—	—	19	—	—	—	—	—
	0.00025	—	2	4	3	2	4	2	—	—	17	—	—	—	—	—
	0.00012	3	5	4	4	2	—	—	—	—	18	—	—	—	—	—
Controls																
Strain PM	10^{-4}	4	3	5	4	2	—	—	—	—	18	0	5	1:640	—	—
Pefloxacin	0.001	1	—	—	—	—	—	—	—	—	1	20	—	—	—	—
Ciprofloxacin	0.001	—	—	—	—	—	—	—	—	—	0	19	—	—	—	—
Roxithromycin	0.001	1	—	—	—	—	—	—	—	—	1	18	—	—	—	—
Amikacin	0.001	—	1	—	—	—	—	—	—	—	—	19	—	—	—	—
Physiological saline	0.05 cm³	1	—	—	—	—	—	—	—	—	1	18	—	—	—	—

Table 13.3 Sensitivity of *C. burnetii* strain PM (*second phase*) to pefloxacin, ciprofloxacin roxithromycin, and amikacin

Type of Antibiotic	Amount of Antibiotic	Chicken Embryos — Days and Daily Mortality after Infection with the Strain (No.)									Total No. of Dead CE	Live CE (No./%)	Guinea Pigs — 9th Day — Days of Fever (No.)	9th Day CF titer	12th Day Days of Fever (No.)	12th Day CF titer
		4	5	6	7	8	9	10	11	12						
Pefloxacin	0.001	—	—	—	—	—	—	—	1	2	3	15 (83.3)	0	1:10	0	1:10
	0.0005	—	—	—	—	—	—	—	2	2	4	15 (78.9)	0	1:10	0	1:10
	0.00025	—	1	2	3	2	1	—	—	—	9	8 (47.0)	1	1:20	1	1:40
	0.00012	—	3	4	3	2	2	—	—	—	14	3 (17.6)	1	1:40	2	1:80
Ciprofloxacin	0.001	—	—	—	—	—	—	1	1	1	3	14 (82.3)	0	1:10	0	1:10
	0.0005	—	—	—	—	2	1	1	1	—	5	14 (73.6)	0	1:10	0	1:10
	0.00025	—	3	4	3	—	—	—	—	—	10	7 (41.1)	1	1:40	2	1:40
	0.00012	3	4	5	3	—	—	—	—	—	15	2 (11.8)	2	1:80	2	1:80
Roxithromycin	0.001	—	—	—	—	—	1	3	2	1	7	12 (63.1)	0	1:10	2	1:10
	0.0005	—	—	2	2	4	3	—	—	—	11	7 (38.9)	2	1:20	2	1:40
	0.00025	—	2	4	3	3	—	—	—	—	12	6 (33.3)	1	1:20	2	1:20
	0.00012	—	4	5	5	2	—	—	—	—	16	2 (11.1)	2	1:40	1	1:80
Amikacin	0.001	—	2	3	2	4	2	3	3	—	19	—	—	—	—	—
	0.0005	—	4	3	2	2	3	3	—	—	17	—	—	—	—	—
	0.00025	2	5	2	3	4	2	—	—	—	18	—	—	—	—	—
	0.00012	3	4	4	3	3	—	—	—	—	17	—	—	—	—	—
Controls																
Strain PM	10^{-4}	3	4	6	2	3	—	—	—	—	18	0	3	1:320	—	—
Pefloxacin	0.001	—	1	—	—	—	—	—	—	—	1	19	—	—	—	—
Ciprofloxacin	0.001	—	—	—	—	—	—	—	—	—	0	20	—	—	—	—
Roxithromycin	0.001	—	1	—	—	—	—	—	—	—	1	18	—	—	—	—
Amikacin	0.001	—	—	—	—	—	—	—	—	—	—	19	—	—	—	—
Physiological saline	0.05 cm³	—	—	—	—	—	—	—	—	—	—	20	—	—	—	—

Upon serological testing, the lowest CF titers (1:10) was established in guinea pigs inoculated with suspensions of YS of CE treated with 0.001, 0.0005, and 0.00025 g of pefloxacin; 0.001 and 0.0005 g of ciprofloxacin; 0.001 and 0.0005 g of roxithromycin. In these animals, the disease was asymptomatic. In contrast to them, the guinea pigs inoculated with the other concentrations used for each of the three antibiotics had higher CF titers to 1:80. By them, the infection manifests itself clinically by an increase in the body temperature within 1 to 2 days (Table 13.2).

The sensitivity of C. burnetii strains in the second phase to the tested antibiotics is illustrated in Table 13.3. The data show that the largest number CE remains alive after treatment with 0.001 and 0.0005 g of pefloxacin, respectively, 83.3% and 78.9%. Close to those rates are the indicators of ciprofloxacin administered at the same concentrations: 82.3%, respectively, 73.6%. The same amount of roxithromycin prevents fewer embryos—63.1% and 38.9%.

At an antibiotic concentration of 0.00025 g, there is a relatively large number of live CE treated with pefloxacin (41.1%) and roxithromycin (33.3%).

Manipulation of embryos with the antibiotics in quantities 0.00012 g had a small protective effect—from 11.1% to 17.6%.

In the tested concentrations of C. burnetii in a second phase (PM strain) was resistant to the action of amikacin (Table 13.3).

The serological investigations of guinea pigs showed that the lowest antibody titers (1:10) were detected after inoculation with the nearest 0.001 g, 0.0005 g of pefloxacin and ciprofloxacin and 0.001 g of roxithromycin. In these animals, no signs of disease were noted, except for two guinea pigs that had temperature rise for two days starting from the 12th day after their infection with a suspension of YS treated with 0.001 g of roxithromycin. In contrast, the guinea pigs inoculated with the suspensions treated with the other concentrations, of the three antibiotics, had higher titers of specific antibodies to 1:80. The disease in these animals was accompanied by hyperthermia, lasting one or more often two days (Table 13.3).

The experiments shown in Tables 13.1–13.3 were accompanied by a number of controls: inoculations of eggs obtained from the same batches with virulent strain PM, with each of the four antibiotics concentration having 0.001 g and separately with saline; analogously, guinea pigs were inoculated with strains and the remaining solutions. The results indicate that the virulent strain untreated with antibiotics leads to 100% lethality of CE and to the development induced by C. burnetii disease in the guinea pigs with a continuous increase in body temperature and a strong serological response (1: 640). The controls with antibiotic solutions and physiological saline lacked similar effects (Tables 13.1–13.3).

The first data received on the sensitivity of C. burnetii in a different phase state to certain antibiotics and quinolones lead to the following conclusion:

a. Coxiella burnetii strains in all three phases, first, transient, and second, have good sensitivity to the quinolones ciprofloxacin and pefloxacin and relatively weak to the macrolide roxithromycin. These strains are resistant to the aminoglycoside amikacin.

b. The most effective is the action of pefloxacin and ciprofloxacin to C. burnetii in its transition phase.

c. Better antibiotic properties against C. burnetii in the first phase have ciprofloxacin, and pefloxacin in the transient and the second phases.

14

Prevention and Control

General prevention or combating Q fever is based on a number of measures and actions that are taken and indicated in the respective country—local, municipal, regional, and national norms and regulations for biosafety—taking into account the characteristics of the pathogen and the disease and the specific conditions in each country.

Undoubtedly, the control, prevention, and the fight against Q fever are directly dependent on the level of scientific knowledge on the properties, referrals, and nature of C. burnetii and the conditions that cause among the representatives of the animal world and man. Identifying the most susceptible to the infection wild and domestic species, spread of disease in them, modes of transmission of the infection, and the complex relationships between hosts and vectors in natural and agricultural foci is the key moment in the development of schemes for prevention and control [30, 46, 111, 228, 353, 509, 525, 560]. Very substantial issues here are the most risky in terms of excretion of the agent physiological periods (birth, lactation, and others) and pathological deviations (abortions and related conditions, mastitis, endometritis, and other clinical forms or asymptomatic latent infections), the role of tick and animal reservoir, the seasonality and resistance of C. burnetii in the environment and others [46, 76, 146, 343, 411, 533, 565]. These features have led to the introduction of actions and measures aimed at reducing excretion and the concentration of the agent in the external environment, the removal and disposal of infected birth products (placentas, fetuses, amniotic fluid, genital secretions), excrement, urine, milk [30, 46, 270, 343, 509, 515, 525]. Basic preventive measure is to fight ticks on animals, especially in areas where respective types of ticks are endemic, and has demonstrated that there exists a tick natural focal reservoir [241]. Regardless of the usefulness of the anti-mite control of ticks, this measure does not always protect animals throughout the season of activity of ticks, so avoiding contact between invaded animals and susceptible individuals (especially during shearing) remains an important moment [241, 296]. Permanent element in the prevention and fight against Q fever is the disinfection after birth or abortion, including disposal of manure with appropriate thermal or chemical exposure before using it for fertilizing [104, 272]. The therapy of animals with antibiotics and the express antibiotic prophylaxis (discussed in the preceding section) are also part of the ways of limiting enzootic and epizootic diseases caused by C. burnetii. American authors [509] stated in 2004 that so far there are no programs for the eradication of Q fever because of rather broad dissemination of the disease among the animals and the difficulties resulting therefrom. Attempts to vaccinate sheep, goats, and cattle remain in the field of research, as it had no practical success [123, 133, 194, 343, 461, 509]. Important preventive measures are the border quarantine, serological testing at import and export of animals and at an annual screening and monitoring of flocks, isolation of clinically ill and serologically positive, restrictive measures in respect of the movement to and from the infected farms and in the assembly of herds [46, 146, 254, 252, 352]. The infected laboratory animals should be severed in rooms with level-3 biosafety [412].

Diagnostic surveys and studies have identified occupational groups at increased risk in terms of Q fever [3, 123, 133, 228, 230, 352, 423, 509, 525, 533]. These include farmers and stockbreeders in contact with sheep, goats, cattle, dogs, and cats; workers in agricultural complexes where there are animals; drivers and loaders trucks serving farms and transporting animals to markets and slaughterhouses; people handling and wrapping meat, milk, leather, and wool; veterinary staff; hunters; persons working in pet stores and zoos; scientists and their collaborators in the laboratories of rickettsiology, virology, microbiology and other areas working with samples of animal strains; staff of vivariums; certain groups of medical workers and researchers who work with blood, sputum, and tissue of infected patients; and servicemen during exercises and operations on the ground. At the risk of infection are also occasional visitors of infected farms and pregnant women. Prevention and efforts to control Q fever should focus primarily on the most vulnerable professional groups and environments [126]. Persons engaged in such activities should be informed about the risks and clinical signs of the disease [134, 133, 412, 525].

Early diagnosis is crucial for successful antibiotic treatment of Q fever [3, 133, 282, 353]. In this regard, serological studies of people by clinical and epidemiological indications as part of a systematic preventive screening are an essential element of the supervision of the disease [352].

As a whole, the prevention of Q fever in humans is based on the control of the disease in the animals [413]. The widespread use of antibiotics and chemotherapeutics in acute and chronic Q fever in humans, including urgent prevention, was discussed earlier in this monograph. Miller [357] states that protection of people from Q fever can be achieved through a combination of common preventive measures to avoid the contact with C. burnetii (protective clothing, hygiene, proper and safe disposal of specified risk materials, disinfection, etc.) with immunological protection against the infection (vaccination). The only available licensed vaccine for Q fever in humans in the world was developed in Australia—GLS Limited of Victoria [126, 129, 357]. The vaccine is prepared of the C. burnetii been cultivated in CE and inactivated with formalin. It is suitable for use in non-immune subjects. Individuals who are immune to C. burnetii should not be vaccinated because of common severe adverse events following vaccination—allergic, local, and systemic reactions [126, 129, 357]. The use of vaccine must be preceded by determining the immune status of the person through skin test and serology on blood sample to detect antibodies against C. burnetii [357]. In Australia, annual detection of 600 cases of the disease in humans is conducted (2001–June 30, 2005), the first of its kind in the world national program of vaccination against Q fever [374]. It covers workers in slaughterhouses, sheep farms, and cattle farms.

Some countries—Australia, Italy, Portugal, USA, and others—have mandatory systems for declaring Q fever, while in many other countries there is no such requirement, and the real incidence remains unclear [126, 185].

In all cases, an effective fight and prevention of Q fever is impossible without joint efforts of the veterinary and medical authorities [2, 37, 38, 140, 413].

The control, combating, and prevention of Q fever are difficult, and this stems from the complex biological nature of the etiological agent, its great resistance, not decrypt pathogenesis, variety of susceptible hosts and vectors in nature and among the domestic animals, a person's sensitivity to infection. To the marked peculiarities should be added a number of other elements: the appearance of new, previously unknown hosts of C. burnetii in the animal world, the mechanisms of transmission of infection, the autonomous cycles, the relationship between natural and agricultural foci [30, 111, 228, 353, 509, 525], the role of tick reservoir the issues at the clinic, the separation of the pathogen from the animals, and the changes in the seasonality [46, 106, 146, 277, 351, 385, 533]. Listed and others not mentioned aspects, factors, and particularities indicate the direct correlation between the level of scientific knowledge on C. burnetii and Q fever and the methods of control, containment, and prevention of disease. Measures in this direction are multidirectional: wide and systematic sero epidemiological investigations of livestock for timely diagnosis and establishing the scope and nosological geography of Q fever; measures to reduce the excretion of the agent in the environment; measures for the removal and disposal of infectious risk products [30, 46, 270, 343, 509, 515], anti-mite control, exterminating, disinfecting [104, 241, 272, 296]; drug therapy and immunization experiments [123, 133, 194, 343, 461, 509]; anti-epizootic restrictions, and border quarantine [46, 114, 146, 254, 352, 525] and others.

It should be emphasized that despite the global spread of Q fever, no single scheme effectively counteracts this zoonosis. French researcher Brouqui [114] is of the opinion that the most effective approach for the prevention of Q fever is a strict application of sanitary and restrictive measures. Detailed analysis of those indisputable realities reveals too complicated picture which highlights a number of problematic issues: uneven level of knowledge of Q fever in animals and humans in various countries on the planet; ambiguous and often unrealistic assessment of the relevance of the problem of veterinary, medical, and social and economic terms in different countries and the resulting incomplete diagnosis, inaccurate information about the spread of disease and lack of purposeful struggle; differences in the degree of scientific diagnostic readiness for demonstration and study of Q fever; ignorance of the spectrum of clinically manifested forms and the availability of less prominent and asymptomatic latent infections in animals; ignorance of diverse but non-specific clinical symptoms in humans, leading to failure to etiologic laboratory tests for Q fever; insufficient studies on treatment with antibiotics and other antibacterial agents, as well as antibiotic prophylaxis of Q fever in animals [122, 156, 277, 380]; lack of practical success of attempts to vaccinate sheep, goats, and cattle with experimental vaccines [123, 133, 194, 277, 343, 461, 509, 596]; restrictions on the use of the world's only licensed vaccine (Australia) in humans, namely only in serologically negative individuals, while in serologically positive C. burnetii individuals is contraindicated because of the frequent and serious post-vaccination events [126, 129, 357]; and the absence in many countries of mandatory notification systems for the disease Q fever in animals and humans [126, 185, 309, 545]. Particular emphasis should be placed on the modern veterinary research, monitoring programs, and schemes for controlling the disease in animals, for which, as mentioned above, the analysis of the literature shows considerably smaller share compared to the developments on Q fever in humans [123, 126, 134, 185, 200, 407, 509, 528].

Obviously, increasing the efficiency in the fighting and prevention of Q fever should be based on updated science programs constructed using a comprehensive approach that will ensure the overcoming of most of these disadvantages.

To answer the need for effective counteraction of Q fever as an important zoonosis with economic, social, health, and strategic character, we have developed a system for monitoring, prevention, and control covering the main necessary actions, measures, and activities designated at the 59 points [597].

14.1 System for Monitoring, Prevention, and Fight against Q Fever

14.1.1 General Preventive Measures and Epidemiological Control

1. All cattle, sheep, and goats imported from abroad are placed in border quarantine for 30 days and are tested serologically for Q fever. Within the country, only serologically negative animals are allowed.
2. Completing of cattle farms, sheep farms, and farms for goats place only with healthy in terms of Q fever animals.
3. Any displacement of cattle, sheep, and goats from one village to another and from one region to another place after prior consultation between the regional authorities of the State Veterinary Service and specifying the epidemiological situation with regard to Q fever based on clinical, epizootiological, and laboratory data, mainly serology.
4. All newcomers in the holdings animals undergo quarantine for 30 days before their introduction into the herd. In this period, serological testing for Q fever takes place.
5. To control the epizootic situation of livestock farms, systematic current and annual serological testings of domestic animals are carried out on personal backyard farms and settlements in terms of Q fever and for the prognosis of the disease.

    5.1. Required prophylactic serological examinations of bulls in the breeding centers and tribal livestock farms.
    5.2. Required virological tests for C. burnetii on imports from abroad of ejaculate from agricultural ruminants.
    5.3. Mandatory serological examinations of all rams from tribal and stock flocks of sheep.
    5.4. Drilling serology on buffaloes, horses, pigs, and poultry.

    5.5. Serological screening for Q fever of ruminants used in experimental research facilities and in schools of veterinary medicine and animal husbandry.

6. The new agricultural foci of Q fever are filing and mapping.
7. Invasive surgery of the abdominal cavity of pregnant sheep, goats, and cows is carried out in the animal medical facility with clinical conditions necessary for these interventions and organization for neutralization of waste organic materials and medical devices.
8. Disclosure filing and mapping of the natural foci of Q fever.

    8.1. Investigations of ticks parasitic on domestic animals from various regions of the country in order to establish of infection with Coxiella burnetii.

    8.2. Tracing the dynamics of species composition, abundance, and seasonality of ticks—reservoirs and carriers of C. burnetii, control on tick populations, and measures to reduce them.

    8.3. Serological and virological tests for infection with C. burnetii of blood sera and other clinical and pathological materials of wild animals and birds.

9. Control and a sharp reduction in the population of stray dogs and cats as an additional reservoir and carrier of C. burnetii.
10. Serological screening for Q fever of dogs and cats from different regions, including major cities.
11. Recording and analyzing the incidence of Q fever in the human populations.
12. A significant expansion of targeted serological examinations of people with pneumonia, including children, with a view to timely disclosure of related C. burnetii etiology and conducting etiological treatment.
13. The persons applying for employment on farms for ruminants must be informed about the potential risks of infection with C. burnetii and get acquainted with clinical signs of Q fever in animals and humans. Training of these persons on ways and means of personal prevention.
14. People with immunodeficiency conditions and diseases of the heart valves and liver should not be allowed to work in risk for Q fever objects.
15. Due to the risk of infection, it is not recommended for pregnant women to work with ruminants, especially in their birth and postnatal periods.
16. Initial and periodic medical examinations and serological tests for Q fever of persons engaged professionally in risk sites—livestock

farms, processing industries, food and raw materials of animal origin, laboratories, and experimental bases with animals and so on.

17. Manufacturing plants processing meat, wool, leather, and other food products and raw materials of animal origin take measures to prevent contamination of the air with dust and aerosols (installing modern ventilation systems, air filtration, and regulation of air flow).

18. Establish a system for prevention and control of Q fever in research and diagnostic laboratories, vivariums, and experimental facilities for animals.

    18.1. Engineering and technical control. Installing a regulated air flow from porches/corridors (zone of minimal contamination) to spaces for work with infectious material (zone of maximum contamination). The air of these premises is not recyclable and is taken out; installing double doors and HEPA filters in rooms, providing laminar boxes with level 2 biological safety serological work and Level 3 when working with C. burnetii strains or with clinical and pathological materials from sick or suspected for Q fever animals. Current and annual engineering control of those facilities in order to prevent the spread of aerosols is carried out.

    18.2. Strict observance of safety rules: use of protective equipment—clothing, masks, goggles, respirators, gloves (latex or vinyl); thoroughly washing of hands after removing gloves and before leaving the laboratory (experimental basis); using liquid antibacterial soap and drying hands with disposable paper; not allowing eating, drinking, smoking, handling contact lenses ophthalmic, cosmetic treatments, touching the mouth and the eyes, mouth pipetting.

    18.3. Strict waste handling biological and medical materials—duly impervious packaging materials, labeling of the same ("Q fever—biological risk"); disposal in incinerators or other authorized incinerators.

    18.4. Disposal of feces and urine of experimentally infected laboratory and other animals as a possible source of contamination with C. burnetii.

    18.5. Decontamination and disinfection of the work place. The tools and materials for reusable undergo autoclaving or chemical disinfection before washing. Disposable clothing is autoclaved before their removal. Garments reusable prior to washing them autoclaved or treated with 0.5% of a hypochlorite solution or a 1:100 solution of lysol. The above solutions were also used for the decontamination of surfaces of equipment.

18.6. Training of staff from laboratories and experimental bases for the risks of infection with C. burnetii, the adequate behavior in workplaces, measures, and rules of personal prevention and protection from contamination of the labor and environment.

19. Mutual information, interaction, and collaboration between veterinary and medical authorities, for controlling, preventing, and fighting Q fever as zoonosis.

20. Training veterinary practitioners and medical staff about the problem Q fever in line with current knowledge and current status of the disease in a given country.

21. Health and educational activities among the population.

14.1.2 Measures to Combat the Disease

1. In cases of proof of Q fever, the holding or its individual farms shall be declared unfortunate and can register.

2. The positive for Q fever cattle is isolated in separate premises for a period of not less than 30 days by providing separate paths, pastures, and watering.

3. All serologically negative in the first test cattle are controlled serologically through 20 days.

4. In case if Q fever is in sheep and goats, then isolate the herd. The other flocks on the holding which are negative in the first study are controlled in 30 days.

5. At the premises of isolated groups of cattle or flocks of small ruminants (serologically positive for Q fever) are separated isolators birth due to a particularly active release of the pathogen from the infected animals in the birth period. Born and aborted animals are kept separately from the others for at least three weeks.

6. Placentas, aborted fetuses, and contaminated litter at birth or abortion destroyed by burning or burying. The materials that are disposed of in incinerators shall be packaged properly and placed signs "Q fever—biological risk." Feces and residues from feed also incinerate or decontaminate. The premises representing isolators' birth,

effluent from them and contaminated objects are disinfected by chlorination/ bleach 50/100 mg/ at exposure 1 hour or 5% chloramine, 5% creolin and creosol, 5% solution of phenol or 5% formalin, heated to 60–70°C. It is also used a 5% sodium hydroxide and 5% solutions of hydrogen peroxide and lysol. At the entrance of the isolators are placed disinfection sites for shoes of the staff.

7. The premises which own the remaining serologically positive animals are disinfected for five days with the above disinfectants. The cleaning of the same daily, manure is composted and undergoes bio-thermal disposal using soon after 6–12 months.

8. Clinical sick and endangered animals in the second half of pregnancy as well as serologically positive animals from herds with active infection with C. burnetii be treated with antibiotics of the tetracycline order parenterally 2 times daily in therapeutic doses for 10–14 days. After a break of one week, the course is repeated. Treated animals were serologically monitored and, if necessary, the treatment is continued. In preparation having depot effect, the frequency of the applications shall be determined by the specified deadline for the maintenance of healing antibiotic concentration.

9. Regular treatment of the animals against ticks—repeatedly every 10 days during the period of their greatest infestation.

10. Carry out systematic mass use of means to combat the rats on farms and private courtyards.

11. The pastures are sanitized and cleaned by eradicating weeds and shrubs, fertilizing, use of chemicals, and so on.

12. Parks and gardens in settlements or in their vicinity, forest and field camps, and similar objects are sanitized and cleaned by regular pruning of shrubs and low trees stem, mowing of lawns, and removing the leaf surface and scrub.

13. The premises is in line with the zoo-hygienic conditions in growing ruminants and provides optimum microclimate, which reduces the possibilities of air-dust infection with C. burnetii and the development of respiratory diseases with mass character.

14. The milk from infected animals shall be decontaminated by boiling for 5 minutes or by prolonged pasteurization (15 min) at a higher temperature (80–90°C). Ordinary pasteurization does not destroy the pathogen. Another way to decontaminate the milk is processing it to cheese, with camping for a term not less than three months.

15. The preparation of butter, cream, and other dairy products from unexploded milk is prohibited.

16. Meat from animals with Q fever is allowed for consumption after neutralization by boiling for 3 hours in pieces weighing up to two kilograms and a thickness of up to 8 cm in open vats or in closed containers at a pressure of 1.5 atmospheres steam for 2½ hours. The internal organs and mammary glands are destroyed by burning or burying.

17. The skins produced in the slaughter of sick animals must be disinfected.

18. The wool and hair of sheep and goats from infected flocks are packaged in opaque containers and sent to the processing plant with an accompanying document indicating its origin of the infected with Q fever herd.

19. Neutralization of wool from sick animals is achieved by washing in warm soap-soda solution (2.5% sodium carbonate and 2.5% soap) at 40 degree celsius for 20 min, followed by disinfection with 2.5% formaldehyde solution at a temperature of 40°C for 20 min. After drying, the wool can be used for processing.

20. In the unfortunate holdings and the farms is prohibited the insertion and the exit of animals to the taking of restrictive measures.

21. Thorough disinfection of vehicles, supplying feed and other materials in the farms with established Q fever.

22. Measures for the prevention of personal staff working in the infected farms. Provides special work clothes—respirators covering the whole face, coat, pants, hat, gloves, rubber boots, masks that after working are disinfected. Clothing is autoclaved in advance, and then washed in a farm itself. Exit from the farms in the settlements with the special clothing is not allowed. The restrooms of the breeders on farms are disinfected regularly.

23. The risk contingencies people are placed under medical observation.

24. The farm is declared a hitch after receiving two consecutive negative serological results in the whole isolated group of animals or herd, and placed under animal health monitoring for another year.

Bibliography

[1] Avakian, A.A., Kulagin, S.M., Kudelin, R.I., and Gulevskaya, S.A. (1970). J. Microbiol., 1, 133.

[2] Alexandrov, E. (1991). Rapid diagnostics, an urgent and specific prevention of some rickettsialpox. D. Sci. Dissertation, VMA – Sofia.

[3] Alexandrov, E., Lesny, M., Samnaliev, M., and Proper, P. (1994). Specific prevention of the rickettsioses. Military Medicine., 3, 26–33.

[4] Alexandrov, E., Propper, P., Dimitrov, D., Lesny, M., Shindov, M., Teoharova, M., Alexandrova, D., Kamarinchev, B., and Samnaliev, M. (1995). Treatment and preventive therapy of the rickettsifses with antibiotics and chemotherapeutics. Infectology, XXXII, 4, 11–15.

[5] Angelov, S. (1951). Reports of Microbiol. Institute, BAS, II, 3, (by Angelov, S., I. Kuyumdjiev S. Gylybov, and Nikolov, P. Q fever in domestic animals. Rep. of Microbiol. Inst., BAS, VIII, 1957), p. 13.

[6] Angelov, C., Kuyumdjiev, Panayotov, P., and Zografski, B. (1953). Rep. of Microbiol. Inst., BAS, IV, 3. (by Angelov, S., I. Kuyumdjiev S. Gylybov, and P. Nikolov. Q fever in domestic animals. Rep. of Microbiol. Inst., BAS, VIII, (1957). 13).

[7] Angelov, C., and Kuyumdjiev, I. (1953). Rep. of Microbiol. Inst, BAS, IV, 23, (by Ognianov D. Q rickettsioses. In: Viral diseases in domestic animals. Zemizdat, Sofia, 1979, 195–200).

[8] Angelov, C., Kuyumdjiev, I., and Galabov, S. (1955). Rep. of Microbiol. Inst., BAS, VI, 29 (by Ognianov D. Q rickettsioses. In: Viral diseases in domestic animals. Zemizdat, Sofia, 1979, 195–200).

[9] Angelov, C, Kuyumdjiev, I., Galabov, S., and Nikolov, P. (1957). Q fever in domestic animals. Rep. of Microbiol. Inst., BAS, VIII, 13–34.

[10] Belchev L., and Pavlov, N. (1977). At histopathology of Q rickettsiosis in sheep. Vet. Science, XIV, 8, 43–48.

[11] Vasilyeva, L.B., Pokorin, O.N., Kekcheeva, N.G., and Yablonskaya, W.A. (1955). Experimental Q rickettsiosis in guinea pigs. ZHMEI, 6, 54–60.

261

[12] Voroshilova, M., Zhevandrova, M., and Balayan, M. (1964). Medicine, Moscow.

[13] Ganchev, N., Abadzhiev, J., Kolev, K., Mihov, M., and Nachev, S. (1969). An epidemic of Q fever in Rousse district. Chronicles of HEI, 9, 100–107.

[14] Ganchev, N., Abadzhiev, J., Genchev, G., and Kolev, V. (1971). Epizootological and epidemiological study of Q fever in Northeastern Bulgaria. Reports of Congress on anthroponoses, Schtrbske Pleso, Czechoslovakia, 6–8 October, 68.

[15] Gening, B. (1969). C. burneti heterogeneous population and breeding analysis. AMS News, USSR, 10, 40.

[16] Gelev I. (1961). Cases of Q fever in sheep and cattle in the region of Ruse station. Updates of TSVIZPB, I, 341–343.

[17] Genchev, D., Kolev, V., and Ognianov, D. (1973). Epidemiological studies on Q fever in sheep. Vet. Science, X 5, 37–42.

[18] Genchev, G.O., and Pavlov, N. (1974). Experimental Q rickettsiosis in guinea pigs and white mice. Vet. Science, XI, 10, 39–47.

[19] Genchev, D., Ognianov, D., and Ganchev, N. (1975). Epizootiological and epidemiological study of Q fever in sheep breeding big complex. Third Nat. Congress of Microbiology, II, 243–248.

[20] Georgieva, G. (1977). Ixodic ticks in Blagoevgrad and Petrich region as carriers of rickettsiae. Chronicles of HEI, 2, 184–187.

[21] Georgieva, G., and Kiossev, B. (1978). Ixodic ticks as vectors of rickettsiae in the settlements of Ruse and Razgrad District with identified foci of Q fever. Chronicles of HEI 8, 180–188.

[22] Georgieva, G., and Filipov, N. (1981). Ecological and geographical characteristics of natural foci of rickettsiae. Sci. theses in Medical Geography "Geographical spread of infectious and parasitic diseases", II, 9–17.

[23] Goldin, R.B., Krstnik, F., and Volkova, L.A. (1965). Papers of Inst. Epidemiol. and Microbiology "Pasteur", 29, 230 (by Serbezov, B., Shishmanov, D., Alexandrov, E., and Novkirishki, V. Rickettsioses in Bulgaria and other Balkan countries' Publisher "H.G. Danov" – Plovdiv, 1973).

[24] Gudima, O.S. (1969). Features of the structural organization of Rickettsia. News of the USSR Academy of Medical Sciences, 10, 35.

[25] Grdevska, B., and Panayotov, A. (1967). Scripta. Sci. Med, 6, 19 (by Serbezov, B., Shishmanov, D., Alexandrov, E., and Novkirishki, V.

Rickettsioses in Bulgaria and other Balkan countries' Publisher "Ch. G. Danov" – Plovdiv, 1973).

[26] Daiter, A.E, Tarasevich, I.V., and Rzhegak, I. (1989). The epidemiology of Q fever. Papers of Inst. Epidemiol. and Microbiology "Pasteur", Leningrad, 66, 5–36.

[27] Drachev, I., and Shishmanov, D. (1958). Military medical work, 19 (by Serbezov, B., Shishmanov, D., Alexandrov, E., and Novkirishki, V. Rickettsioses in Bulgaria and other Balkan. Publisher "Ch. G. Danov" – Plovdiv, 1973).

[28] Zhokovsky, N., and Minchev, G. (1975). Study the spread of Q fever in Vratsa District. Sci. Works of NRC – Vratsa vol. II, Sofia, 61–67.

[29] Zografski, B., Doynov, M., and Beloev, J. (1959). Hyg. Epidemiol., Microbiol., 1, 29 (in Serbezov, C., D. Shishmanov E. Alexandrov, V. Novkirishki. Rickettsifses in Bulgaria and other Balkan countries. Publisher "Ch. G. Danov" – Plovdiv, 1973).

[30] Zdrodovsky, P.F., and Golinevich, E.M. The doctrine of Rickettsiae and rikketsioses, Medicine, Moscow, 1972.

[31] Kambaratov, P.I., Kudelina, R.I., Gavrilov, N.A., and Lvova, K.S. (1970). J. Microbiol., 9, 87.

[32] Kekcheeva, N.G. (1955). Experimental chemotherapy of Q fever ZHMEI, 6, 60–65.

[33] Kechkeeva, N.D., and Kokorin, I.N. (1955). Bull. Exp. Biol. and medicine 6, 46 (in Kokorin, I.N. An attempt for morphological study of processes in infections and immunity at experimental brucellosis and rickettsioses. D. Sci. Dissertation, Moscow, 1957.

[34] Kuyumdjiev, I. (1957). Q fever. Theses of the reports of the Conference on some zoonoses and diseases with nature foci, Plovdiv, 15–17.

[35] Martinov, S.P., and Popov, G. (1980). Chlamydia and chlamydial infections (overview). NAPS-CSTI, Sofia, 88 pp.

[36] Mitov, A., Shindarov, L., and Serbezov, V. (1964). Q fever in Bulgaria, ZHMEI, 8, 101–106.

[37] National program for prevention and control of the tick-borne transmissible infections in Bulgaria. Decision No. 18/16.01.2004 of the Council of Ministers.

[38] National interdepartmental expert council for control of the organization, coordination and implementation of the National Programme for Prevention and Combating the tick-borne transmissible infections in Bulgaria. MH – Order No. RD-09-343/02.06.2004.

[39] Novkirishki, C., Boyadzhiyan, H., Kanyovska, E., Hristov, H., Gancheva G., Obretenova, J., Tsvetkova, I., Angelov, P., and Rounevska, D. (1994). Studies on the epidemic of Q fever in the area of the town Knezha. Infectology, XXXI, 3, 16–19.

[40] Nikolov P. (1961). Q rickettsioses. Rep. of Microbiotol. Inst., BAS, XIII, 89.

[41] Nikolov, P. (1961). Comprehensive studies of natural outbreaks in the Strandja Mountain, Rep. of Microbiotol. Inst., BAS, XII, 73.

[42] Nikolov P. (1962). Natural infection foci in Petrich and Gotse Delchev, Proceedings of Division biol. science, Sofia, BAS, 51.

[43] Ognianov D. (1962). Studies on an epizootic of Q fever in goats. Reports of the Central Research Veterinary Institute of Virology, Sofia, 1, 159–163.

[44] Ognianov, D., Genchev, D., and Kolev, V. (1971). Studies on Q fever in sheep. Veterinary sciences, VIII 2, 17–22.

[45] Ognianov, D., Pavlov, N., and Genchev, D. (1975). Coxiella burnettii as the etiologic agent in respiratory disease in lambs. Third National Congress of Microbiology, Sofia, II, 235–237.

[46] Ognianov, D. (1979). Q-rickettsiosis. In: Viral diseases in domestic animals /H. Haralampiev, Dilovski, M., Eds./, Zemizdat, Sofia, 195–200.

[47] Pandarov, S. (1968). Q fever in sheep. Vet. sbirka, XV, 1, 13–14.

[48] Pandarov, S., and Dimitrov, A. (1969). Clinical and post-mortem studies in sheep and laboratory animals infected with C. burnetii. Veterinary sciences, 3, 53–61.

[49] Pandarov, S. (1970). Studies on Q fever in sheep. First National Conference on zoonoses, Sofia, 21.

[50] Pautov, V.N., and Igumenov, A.M. (1968). Rickettsia Biology. Medicine, Moscow, 25.

[51] Pautov, V.N. (1973). Rickettsioses. ZHMEI, 3, 74–77.

[52] Petrov, S., C. Serbezov N. Nenov, and Alexandrov, E. (1969). Internal Medicine, 8, 437. (in Serbezov, C., D. Shishmanov, E. Alexandrov, V. Novkirishki. Rickettsioses in Bulgaria and other Balkan countries. "Ch. G. Danov" – Plovdiv, 1973).

[53] Semenov, B.F. (1965). Diagnosis of rickettsiae in Lab. diagnostics viral and rickettsial diseases, ZHMEI, 6, Moscow, 209–211.

[54] Semerdzhiev, B., and Ognianov, D. (1969). Neorickettsioses of domestic animals. Zemizdat, Sofia, 252 pp.

[55] Serbezov, V. (1959). Hygiene, Epidemiology and Microbiology, 1, 40–47. (by Serbezov, C., D. Shishmanov E. Alexandrov, V. Novkirishki. Rickettsioses in Bulgaria and other Balkan countries. "Ch. G. Danov" – Plovdiv, 1973).

[56] Serbezov, V. (1961). Hygiene, 6, 47 (in Serbezov, C., D. Shishmanov E. Alexandrov, and Novkirishki, V. Rickettsioses in Bulgaria and other Balkan countries. Publisher "Ch. G. Danov" – Plovdiv, 1973).

[57] Serbezov, V. (1965). Q fever in Bulgaria. PhD thesis, ISUL, Sofia 1961.

[58] Serbezov, C., D. Mitov Kaprelyan D., and Galabov, K. Epidemiol., Microbiol. and Infect. Dis., 1, 40.

[59] Sidorov, V., and Kudelina, R. (1973). ZHMEI 11, 118. (in; Pandarov S., K. Akabalieva. Q fever. Zemizdat, (1985). 127 pp.)

[60] Sprostranov, V.I. (1957). Contemporary Medicine, 4, 93.

[61] Fedorova, N.I. (1968). Epidemiology and Prevention Q rickettsiosis. D. Sci. Dissertation. Moscow.

[62] Tsvetanov L. (1957). Hygiene, epidemiology, microbiology, 6, 41.

[63] Tsonchev, I., and Karacholov, I. (1955). Contemporary medicine, 4, 3. (by Serbezov, C., D. Shishmanov E. Alexandrov, V. Novkirishki. Rickettsioses in Bulgaria and other Balkan countries. "Ch. G. Danov" – Plovdiv, 1973).

[64] Shindarov, L., Dimov, I., Glindzhev, I., Genov, D., and Solakov, P. (1957). Hygiene, epidemiology, microbiology, 6, 35.

[65] Shindarov L. (1958). Acta Virol., 1, 62 (by Alexandrov, E. Rapid diagnostics, an urgent and specific prevention of some rickettsioses. D. Sc. Dis., Sofia, 1991).

[66] Shindarov, L., Mitov, A., Serbezov, V., Mitov, D., Matova, E., and Ivanov, M. (1959). Contemporary medicine, 2, 3. (by Ognianov D. Studies on a enzootics of Q fever in goats. Rep. Central. Res. Vet. Inst. of Virology, (1962). 159–163).

[67] Yarmin, P. (1993). Study on the natural foci and epidemiology of Q fever. Proc. Conference "40 years RVS Vidin", 115–118.

[68] Swayne, D.E., Glisson, J.R., Jackwood, M.W., Pearson, J.E., and Reed, W.M. (eds.) (1998). A laboratory manual for the isolation and identification of avian pathogens, 4th Edition, Am. Assoc. avian pathologists, Ins., Kennett. Sq., Pennsylvania, USA.

[69] Aitken, I.D., Bögel, K., Gracea, E., Edlinger, E., Houwers, D., Krauss, H., Rady, M., Rehacek, J., Schiefer, H.G., Schmeer, N., Tarasevich, I.V., and Tringali, G. (1987). Q-fever in Europe: Current aspects of aetiology,

epidemiology, human infection, diagnosis and control. Infection, 15, 323–327.

[70] Akporiaye, E.T., Rowatt, J.D., Aragon, A.A., and Baca, O.G. (1983). Lysosomal response of a murine macrophage—like cell line persistently infected with C. burnetii. Infect. Immun., 40, 1155–1162.

[71] Aldomi, F.M.M., and Wilsmore, A.J., and Saft, S.H. (1998). Q-fever and abortion in sheep and goats in Jordan. Pakistan Vet. J., 18 (1), 43.

[72] Alexandrov, E.G., Teoharova, M., Kamarinchev, B.D., Mitov, D.G., and Bogdanov, N.R. (1999). Etiological role of C. burnetii and SFG rickettsiae in atypical pneumonia. In: Rickettsiae and rickettsial diseases at the turn of the third millenium (D. Raoult, Brouqui, P., eds.), Elsevier, 282–284.

[73] Amano, K.I., Williams, J.C., McCaul, T.F., and Peacock, M.G. (1984). Biochemical and immunological properties of C. burnetii cell wall and peptidoglycan—protein complex fractions. J. Bacteriol., 160, 982–988.

[74] Amano, K.I., and Williams, J.C. (1984). Chemical and immunological characterization of lipopolysaccharides from phase I and II C. burnetii. J. Bacteriol., 160, 994–1002.

[75] Amano, K.I., Williams, J.C., Missler, S.R., and Reinhold, V.N. (1987). Structure and biological relationships of C. burnetii lypopolysaccharides. J. Biol. Chem., 262, 4740–4747.

[76] Anacker, R.L., Fukushi, K., Pickens, E.G., and Lackman, D.B. (1964). J. Bacteriol., 88, 1130.(по Сербезов, В., Д. Шишманов, Е. Александров, В. Новкиришки. Рикетсиозите в България и другите балкански страни., „Хр. Г. Данов” —Пловдив, 1973).

[77] Anderson, B.E., Dawson, J.E., Jones, D.C., and Wilson, K.H. (1991). Erlichia chaffeensis, a new species associated with human erlichiosis. J. Clin. Microbiol., 29, 2838–2842.

[78] Anusz, Z., Řeháček, J., Knap, J., Krauss, H., Ciecierski, H., A. Platt-Samoraj, Wodecki, W., and Lewicki, P. (1992). Epizootiology and epidemiology of Q fever in cattle and man in the Zulawy region. Acta Agr. Tech. Oltensis, Veterinaria, 20, 35–45.

[79] Anusz, Z. (1993). Analysis of the epizootiological and epidemiological aspects of Q fever in Poland and in the world in 1956–1991. Acta Acad. Agr. Tech. Olstensis, Veterinaria, 21, 3–19.

[80] Apostol, B.L., W.C. Black IV, Miller, B.R., Reiter, P., and Beaty, B.J. Estimation of family members at an oviposition site using RAPD-PCR markers: application to the moskito Aedes aegypti. Theor. Appl. Cenetics, 85, 991–1000.

[81] Armengaud, A., Kessalis, N., Decenclos, J.C., Maillot, E., Brousse, P., Brougui, P., Tixier-Dupont, H., Raoult, D., Provensal, P., and Ovadia, Y. (1997). Urban outbreak of Q fever, Briancon, France, March to June 1996. Eurosurveillance, 2 (2), 12–13.

[82] Atzopien, E., W. Baumgärtner, Artelt, A., and Thiele, D. (1994). Valvular endocarditis occurs as a part of a disseminated C. burnetii infection in immunocompromised BALB/cJ (H2^d) mice infected with Nine Mile isolate of C. burnetii. J. Infect. Dis., 170, 223–226.

[83] Ayres, J.G., Smith, E.G., and Flint, N. (1996). Protracted fatique and debility after acute Q fever. Lancet, 347, 978–979.

[84] Ayres, J.G., Flint, N., Smith, E.G., Tunniclife, W.S., Fletcher, T.J., Hammond, K., Ward, D., and Marmion, B.P. (1998). Post-infection fatique syndrome following Q fever. Q.J. Med., 91, 105–123.

[85] Babudieri, B. (1959). Q fever: a zoonosis. Adv. Vet. Sci., 5, 81.

[86] Baca, O.G., Akporiaye, T., Aragon, A.S., Martinez, I.L., Robles, M.V., and Warner, N.L. (1981). Fate of phase I and phase II C. burnetii in several macrophage-like tumor cell lines. Infect. Immun., 33, 258–266.

[87] Baca, O.G., and Paretsky, D. (1983). Q fever and C. burnetii: a model for host-parasite in—teractions. Microbiol. Rev., 47, 127–149.

[88] Baca, O.G., Scott, T.O., Akporiaye, T., De Blassie, R., and Crissman, H.A. (1985). Cell cycle distribution patterns and generation times of L 929 fibroblasts cells persistently infected with C. burnetii. Infect. Immun., 47, 366–369.

[89] Banerjee-Bhatnagar, N., Bolt, C.R., and Williams, J.C. (1996). Pore-forming activity of C. burnetii outher membrane protein oligomer compised of 29.5—and 31-Kda polypeptides. Inhibition of porin activity by monoclonal antibodies 4E8 and 4D6. Ann. N.Y. Acad. Sci., 791, 378–401.

[90] Bartlett, J.G. (2000). Questions about Q fever. Medicina (Baltimore), 79 (2), Mar, 124–125.

[91] Beaman, M.H., and Hung, J. (1989). Pericarditis associated with tick-borne Q fever. Aust. N.Z. J. Med., 19, 254–256.

[92] Behymer, D., Ruppaner, D., Riemann, H., Biberstein, E., and Franti, C. (1977). Folia Veterinaria Latina, 7, 64–70. (по Пандъров, С., К. Акабалиева. Ку-треска. Земиздат, 1985).

[93] Benson, W.W., Brack, D., and Mather, J. (1963). Serological analysis of a penitentiary group using raw milk from a Q fever infected herd. Public Health Rep., 78, 707–710.

[94] Bental, T., Fejgin, M., Keysary, A., Rzotkiewicz, S., Oron, C., Nachum, R., Beyth, Y., and Lang, R. (1995). Chronic Q fever of pregnancy presenting as C. burnetii placentitis: succesful outcome following therapy with erythromycin and rifampin. Clin. Infect. Dis., 21, 1318–1321.

[95] Berkovich, M., Aladjem, M., Beer, S., and Cohar, K. (1985). A fatal case of Q fever hepatitis in a child. Helv. Paediatr. Acta, 40, 87–91.

[96] Berri, M., Laroucau, K., and Rodolakis, A. (2000). The detection of C. burnetii from ovine genital swabs, milk and fecal samples by use of a single touchdown PCR. Vet. Microbiol., 72, 285–293.

[97] Berri, M., Souriau, A., Crosby, M., Crochet, D., Lechopier, P., and Rodolakis, A. (2001). Relationships between the shedding of C. burnetii, clinical signs and serological responses of 34 sheep. Vet. Rec., 148, 502–505.

[98] Bieling, R. (1950). Balkangrippe das Q Fieber der alten Welt. Beiträge zur Hyg. & Epidemiol. H., 5.

[99] Biosafety in Biomedical and Microbiological laboratories, Section VII—E, Rickettsial agents. Coxiella burnetii, (1999). http://bmbl.od. nih.gov/rickettsia.htm

[100] Blair, P.J., Schoeler, G.B., Moron, C., Anaya, E., Caceda, R., Cespedes, M., Cruz, C., Felices, V., Guevara, C., Huaman, A., Luckett, R., Mendoza, L., Richards, A., Rios, Z., Sumner, J.W., Villaseca, P., and Olson, J.G. (2004). Evidence of rickettsial and leptospiral infections in Andean Northern Peru. Am J. Trop. Med. Hyg., 70, 357–363.

[101] Blondeau, J.M., Yates, L., Martin, R., Marrie, T., Ukoli, P., and Thomas, A. (1990). Q fever in Socoto, Nigeria. Ann. N.Y. Acad. Sci., 590, 281–282.

[102] Blondeau, J.M., Williams, J.C., and Marrie, T.J. (1990). The immune response to phase I and phase II antigen as measured by Western immunoblotting. Ann. N.Y. Acad. Sci., 590, 187–202.

[103] Boni, M., Davoust, B., H. Tissot-Dupond, and Raoult, D. Vet. Microbiol., 64, (1), 1998 Nov, 1–5.

[104] Böhm, R., and Strauch, D. Desinfection im stall—weniger Krankheiten, mehr Leistung. Bonn: Gesellschaft fur Druckabwicklung mbH, 1996.

[105] Bouvery, N.A., Souriau, A., Lechoper, P., and Rodolakis, A. (2003). Experimental C. burnetii infection in pregnant goats: excretion routes. Vet. Res., 34, 423–433.

[106] Bouvery, N.A., and Rodolakis, A. (2005). Is Q Fever an emerging or re-emerging zoonosis? Vet. Res., 36, 327–349.

[107] Brennan, R.E., and Samuel, J.E. (2003). Evaluation of C. burnetii antibiotic susceptibilities by Real-Time PCR Assay. J. Clin. Microbiol., 41, 1869–1874.

[108] Brezina, R., and Rehaček, J. (1961). Acta virol., 5, 250. (по Пандъров, С., К. Акабалиева. Ку-треска, Земиздат, 1985).

[109] Brezina, K., and Kazar, I. (1963). Actavirol., 7, 746. (по Сербезов, В., Д. Шишманов, Е. Александров, В. Новкиришки. Рикетсиозите в България и другите балкански страни., „Хр. Г. Данов" – Пловдив, 1973).

[110] Brezina, R. (1976). Phase variation phenomenon in Coxiella burnetii. In: Rickettsia and rickettsial disease (J. Kazar, Ormsbee, R.A., I. Tarasevitch). Veds. Edit, Bratislava, 221–235.

[111] Brooks, D.L., Ermel, R.W., Franti, C.E., Ruppanner, R., Behymer, D.E., Williams, J.C., and Stephenson, J.C. (1986). Q fever vaccination of sheep: challenge of immunity in ewes. Am. J. Vet. Res., 47, 1235–1238.

[112] Brouqui, P.H., H. Tissot-Dupond, Drancourt, M., Berland, Y., Etienne, J., Leport, C., Coldstein, F., Massip, P., Micoud, M., Bertrand, A., and Raoult, D. (1993). Epidemiologic and clinical features of chronic Q fever: 92 cases from France (1982–1990). Arch. Intern. Med., 153, 642–648.

[113] Brouqui, P., Dumler, J.S., and Raoult, D. (1994). Immunohistological demonstration of Coxiella burnetii in the valves of patients with Q fever endocarditis. Am. J. Med., 97, 451–458.

[114] Brouqui, P. (1997). Tropical animal and human rickettsial infections. Med. Trop. (Mars), 57 (3 Suppl.), 23–27.

[115] Buhariwalla, F., Cann, B., and Marrie, T.J. (1996). A dog related outbreak of Q fever. Clin. Infect. Dis., 23, 753–755.

[116] Burgdorfer, W. (1970). Hemolymph test. A technique for detection of rickettsiae in tick. Am. J. Trop. Med. Hyg., 19, 1010–1014.

[117] Burnet, F.M., and Freeman, M. (1937). Experimental studies on the virus of Q fever. Med. J. Aust., 299–305.

[118] Byrne, W.R. (1997). Q-fever. In: Medical aspects of Chemical and Biological Warfare. (R. Zajtchuk, Bellamy, R.F., eds.). Washington, DC Office of the Surgeon General US Dept. of the Army, 523–537.

[119] Camacho, M.T., Outschoorn, I., Kovacova, E., and Tellez, A. (1996). IgA subclasses in Q-fever. In: Rickettsiae and Rickettsial diseases (J. Kazar, Toman, R., eds.), Bratislava, VEDA, 455–458.

[120] Caminopetros, I. (1948). La Q-fever en Grece. Le lait sotse de infection etc. Ann. Parasitologie, 23, 107.

[121] Caminopetros, I. (1949). La bronchopneumonia epidemique hiverno-printaniere. Ann. Institute Pasteur, 77, 6.

[122] Can. Lab. Centre for Disease Control. Material safety data sheet—Coxiella burnetii. Dec. 2002, http://www.hc-ss.gc.ca/pphb-0dgspsp/msds43e.html

[123] Canadian centre for Occupational Health & Safety. OSH Answers: Q-fever. (2005). http://www.ccohs.ca/oshanswers/diseases/qfever.html

[124] Campbell, F.R.S. (1994). Pathogenesis and pathology of the complex rickettsial infections. Vet. Bull., 64, (1), 2–24.

[125] Capponi, M. (1974). Diagnostic des Rickettsiales au laboratoire. Maloine, Paris, France, 89–94.

[126] CDC—Viral and Rickettsial Zoonoses branch. Q-fever, (2005). http://www.cdc.gov/ncidod/dvrd/qfever/

[127] Cekanać, R., Lukac, V., and Coveljic, M. An epidemic of Q fever in a unit of the Yugoslav Army during war conditions. Vojnosanit. Pregl., 59 (2), 2002 Mar–Apr., 157–160.

[128] Centers for Disease Control, Nat. Inst. of Health. Tick borne diseases—Q-fever, (2002). http://www.stopticks.org/ticks/qfever.asp

[129] Centers for disease control and prevention. Viral and Rickettsial Zooneses branch, Atlanta, GA, USA. Q-fever, 5 pp, (2003). http://www.cdc.gov/ncidod/dvrd/qfever

[130] Chen, S.-Y., Hoover, T.A., Thompson, H.A., and Williams, J.C. (1990). Characterization of the origin of DNA replication of the C. burnetii chromosome. Ann. N.Y. Acad. Sci., 590, 491–503.

[131] Chen, S.-Y., Vodkin, M., Thompson, H.A., and Williams, J.C. (1990). Isolated Coxiella burnetii synthesizes DNA during acid activation in the absense of host cells. J. Gen. Microbiol., 136, 89–96.

[132] Chevalier, P., Vandenesch, F., Brouqui, P., Kirkorian, G., Tabib, A., Etienne, J., Raoult, D., Loire, R., and Touboul, P. (1997). Fulminant myocardial failure in a previously healthy young man. Circulation, 95, 1654–1657.

[133] Child and Youth Health. Q-fever, (2005). http://www.cyh.com/health topicsdetails.aspxp=114&mp=303&id=19...

[134] Chin, J. Control of Communicable Dis. Manual, Seventeenth Edition, Am. Publ. Health Assoc., 2000.

[135] Cho, S.-N. (1996). Prevalence of antibodies to Coxiella burnetii among residents in Korea. The Korea-Bulgaria int. Seminar on Q-fever and Chlamydiosis, Apr. 15–24, Anyang, R. of Korea, 6.

[136] Choiniere, Y., and Munroe, J. Farm workers' health problems related to air quality inside livestock barns. Ontorio, Canada: Canadian Government Publishing, Health Canada, 2001.

[137] Chopra, I., and Roberts, M. (2001). Tetracycline antibiotics: mode of action, applications, molecular biology and Epidemiology of Bacterial Resistance. Microbiol. Mol. Biol. Rev., 65, 232–260.

[138] Cicalini, S., Forcina, G., and De Rosa, F.G. (2001). Infective endocarditis in patients with human immunodeficiency virus infections. J. Infect. Dis., 42, 267–271.

[139] Colbeck, I. Physical and chemical properties of aerosols. London: Blackie Academic Press, 1998.

[140] Coleman, T.J. (2000). The public health laboratory service (PHLS) and its role in the control of zoonotic diseases. Acta trop., 1, 71–75.

[141] Comer, J.A., Paddock, C.D., and Childs, J.E. (2001). Urban zoonoses caused by Bartonella, Coxiella, Ehrlichia and Rickettsia species. Vector borne zoonotic dis., 1, 91–118.

[142] Combiesco, D., Dumitresco, N., Botez, V., Sturdza, N., and Zornea, O. Roum. Pathol. Exp. Microbiol., 15, 242. (По Combiescu, D., N. Dumitrescu. Noicesari de tifos pulmonar (febre Q) ivite intre lugratorii unei ferme. Bull. Stiint. Acad. R.P. R., 2, (1950). 733).

[143] Commission on Acute Respiratory Diseases. (1946). Am. J. Hyg., 44, 88.

[144] Commission on Acute Respiratory Diseases. (1946). Am. J. Hyg., 44, 100.

[145] Commission on Acute Respiratory Diseases. (1946). Am. J. Hyg., 44, 103.

[146] Control of Communicable Diseases. (2000). Edited by J. Clin. Am. Publ. Health Assoc., 407–411.

[147] Cottalorda, J., Jouve, J.L., Bollini, G., Touzet, P., Poujol, A., Kelberine, F., and Raoult, D. (1995). Osteoarticular infection due to Coxiella burnetii in children. J. Pediatr. Orthop., 4, 219–221.

[148] Cowley, R., Fernandes, F., Freemantle, W., and Rutter, D. (1992). Enzyme immuno-assay for Q fever: comparision with complement fixation and immunofluorescence test and dot immunoblotting. J. Clin. Microbiol., 30, 2451–2455.

[149] Cox, H.R. (1938). A filter-passing infectious agent isolated from ticks. III. Description of organism and cultivation experiments. Public Health Rep., 53, 2270–2276.

[150] Cox, H.R., and Bell, E.J. (1939). The cultivation of Rickettsia diaporica in tissue culture and in the tissues of developing chicken embryos. Public Health Rep., 54, 2171–2175.

[151] Dackau, T. (1993). Development of capture ELISA for detection of C. burnetii in milk samples—a possible alternative to the guinea pig inoculation test. Berl. Münch. Tierärztl. Woch., 106 (3), 87–90.

[152] Dalton, A.J. (1955). Ein Chrom-Osmium Fixative für die Electronen-microscopie. Anatom. Rec., 121, 281.

[153] Daoust, P.J., and Perry, R. (1989). Coxiellosis in a kitten. Can. Vet. J., 30, 434.

[154] Dasch, G.A., Weiss, E., and Williams, J.C. (1990). Antigenic properties of Erlichiae and other Rickettsiaceae. Current topics in Vet. Med. & Animal Science, 54, 35–58.

[155] Davis, G.E., and Cox, H.R. (1938). A filter-passing infectious agent isolated from ticks. I. Isolation from Dermacentor andersoni, reactions in animals, and filtration experiments. Public Health Rep., 53, 2259–2261.

[156] De la Concha-Bermejillo, A., Kasari, E.M., Russell, K.E., Cron, L.E., Browder, E.J., Callicott, R., and Ermell, R.W. Q-fever: An overview. US Anim. Health Assoc., 4 Dec 2002, http://www.usaha.org/speeches/spee ch01/s01conch.htm

[157] Dedie, K., Bockemühl, J., Kühn, H., Volmer, K.-J. and Weinke, T. (1993). Bakt. Zoonozen bei Tier und Mensch. Epidemiol., Pathol., Klinik, Diagnostic und Bekämpfung. Sttutgart, Germany, Ferdinand Enke Verlag.

[158] Dellacasagrande, J., Capo, C., Raoult, D., Mege, J.-L. (1999). IFN-γ-mediated control of C. burnetii survival in monocytes: the role of cell apoptosis and TNF[1]. J. Immunol., 162, 2259–2265.

[159] Dellacasagrande, J., Chigo, C., O.S. Machergui-El, Hammami, T., Toman, R., and Raoult, D. (2000). Alpha vbeta 3 Integrin and Bacterial Lipopolysaccharide are involved in C. burnetii—stimulated Production of tumor necrosis factor by human monocytes. Infect. Immun., 68, 5673–5678.

[160] Dellacasagrande, J., Moulin, P.A., Guilianelli, C., Capo, C., Raoult, D., Grau, G.E., and Mege. J.-L. (2000). Reduced transendothelial migration of monocytes infected by Coxiella burnetii. Infect. Immun., 68, 3784–3786.

[161] Dellacasagrande, J., Ghigo, E., Raoult, D., Capo, C., and Mege, J.-L. (2002). IFN-γ-induced adoptosis and microbicidal activity in

monocytes harboring the intracellular bacterium C. burnetii require membrane INF and homotypic cell adherence. J. Immunol., 169, 6309–6315.

[162] Denning, H. (1942). Wien. Med. Wschr., 92, 335.

[163] Department of the Army. US Army activity in the US Biological Warfare Program. Vol. 2, Washington DC, HQ, DA, 24 Feb 1977, D 1–2, Appendix 4, pp, E 4-1. Unclassified.

[164] Derrick, E.H. (1937). "Q" fever, new entity: clinical features, diagnosis and laboratory investigation. Med. J. Aust., 2, 281–299.

[165] Derrick, E.H. (1973). The course of infection with Coxiella burnetii. Med. J. Aust., 1, 1051–1057.

[166] Dindinaud, G., Aigius, G., Burucoa, C., Senet, J.M., Desahyes, M., and Magnin, G. (1991). Q fever and fetal death in utero. Two cases. J. Gynecol. Obstet. Biol. Reprod. (Paris), 20, 969–972.

[167] Dolcé, P., M.-J. Bélanger, and Tumanowicz, K. Coxiella burnetii seroprevalence of shepherds and their flocks in the lower Saint-Lawrense River region of Quebec. Canad. J. Infect. Dis. & Med. Microbiol., 14 (2), March–April 2003, 97–102.

[168] Doller, G., Doller, P.C., and Gerth, H.J. (1984). Early diagnosis of Q fever: detection of immunoglobulin M by radioimmuassay and enzyme immunoassay. Eur. J. Clin. Microbiol. Infect. Dis., 3, 550–553.

[169] Dorko, E., and Čislakova, L. (2004). Q-fever—incidence in the Slovak and Czech Republics. Slovensky Vet. Časopis, 29 (5), 26–28.

[170] Drevets, D.A., Leenen, P.J.M., and Greenfield, R.A.O. (2004). Invasion of the central nervous system by intracellular bacteria. Clin. Microbiol. Rev., 17, 323–347.

[171] Drevets, D.A., Leenen, P.J.M., and Greenfield, R.A.O. (2004). Confusion and lethargy in a 48 year old man. Postgrad. Med. J. 80, 244–245.

[172] Dupuis, G., Peter, O., Petroni, D., and Petite, J. (1985). Clinical aspects observed during an epidemic of 415 cases of Q fever. An important outbreak of human Q fever in a Swiss alpine valley. Schweiz. Med. Wochenschr., 115, 814–818.

[173] Dupuis, G., Petite, J., and Olivier, P.V. (1987). An important outbreak of human Q fever in a Swiss alpine valley. Int. J. Epidemiol., 16, 282–287.

[174] Dyer, R.E. (1938). A filter-passing infectious agent isolated from ticks, IV: Human infection. Public Health Rep., 53, 2277–2283.

[175] Dyer, R.E. (1939). Similarity of Australian Q-fever and a disease caused by an infectious agent isolated from ticks in Montana. Public Health Rep., 54, 1229.

[176] Echaniz, A., Miguez, P., and Llinares, P. F. (1992). Diz-Lois. Recurrent pericarditis caused by Q fever. Rev. Clin. Esp., 191, 170–171.

[177] Ejercito Cal, L., Htwe, K.K., Taki, M., Inoshima, Y., Condo, T., Kano, C., Abe, S., Shirota, K., Sugimoto, K., Yamaguchi, T., Fukushi, H., Minamoto, N., Kihjo, T., Isogai, E., and Hirai, K. (1993). Serological evidence of Coxiella burnetii infection in wild animals in Japan. J. of Wildlife Dis., 29 (3), 481–484.

[178] Eklund, C.M., Parker, R.R., and Lakman, D.B. (1947). A case of Q fever probably contracted by exposure to ticks in nature. Publ. Health Rep., 62, 1413–1416.

[179] Elis, M.E., Smith, C.C., and Moffat, M.A. (1983). Chronic and fatal Q-fever infection: a review of 16 patients seen in North-East Scotland (1967–1980). Q.J. Med., 52, 54–66.

[180] Elliot, T.S.J., Foweraker, J., Gould, F.K., Perry, J.D., and Sandoe, J.A.T. (2004). Guidelines for the antibiotic treatment of endocarditis in adults: report of the Working Party of the British society for antimicrobial Chemotherapy. J. Antimicrob. Chemother., 54, 971–981.

[181] Enright, J.B., Longhurst, W.M., Franti, C.E., Behymer, D.E., Duston, V.J., and Wright, M.E. (1971). Am. J. Epidemiol., 94, 72–78.

[182] Eremeeva, M., Dasch, G.A., and Silverman, D.J. Quantitative analyses of variations in the injury of endothelial cells elicited by 11 isolates of R. rickettsii. Clin. Diagn. Lab. Immunol., 8 (4), July 2001, 788–796.

[183] Erlich, H.A., Gelfand, D.H., and Saiki, R.K. (1988). Specific DNA amplification. Nature (London), 461–462.

[184] Erlich, H.A. (1989). PCR technology: principles and applications for DNA amplification. Stockton Press, New York.

[185] Eurosurveillance. Q-fever in Europe. Eur. Communicable disease bull., 2 (2), Feb 1997, 13–15.

[186] Feinstein, M., Yesner, R., and Marks, J.L. (1946). Am. J. Hyg., 44, 72.

[187] Felsenstein, J. (1993). PHYLIP (Phylogeny inference package) version 3.5c. Dept. of Genetics, Univ. of Washington, Seattle.

[188] Fergusson, R.J., and Shaw, T.R.D., Kitchin, A.H., Mattews, M.M., Inglis, J.M., and Peutherer, J.F. (1985). Subclinical chronic Q fever. Q.J. Med., 57, 669–676.

[189] Fenollar, F., Fournier, P.E., Carrieri, M.P., Habib, G., Messana, and Raoult, D. (2001). Risks factors and prevention of Q fever endocarditis. Clin. Infect. Dis., 33, 312–316.

[190] Ferrante, M., and Dolan, M.J. (1993). Q fever meningoencephalitis in a soldier returning from the Persian Gulf War. Clin. Infect. Dis., 16, 489–496.

[191] Field, P.R., Hunt, J.G., and Murphy, A.M. (1983). Detection and persistence of specific IgM antibody to C. burnetii by enzyme-linked immunosorbent assay: a comparision with immunofluorescence and complement fixation tests. J. Infect. Dis., 148, 477–487.

[192] Field, P.E., Mitchell, J.L., and Santiago, A. (2000). Comparision of a commercial ELISA with IF and CFT for detection of C. burnetii (Q fever) immunoglobulins. J. Clin. Microbiol, 4, 1645–1647.

[193] Fiset, P. (1959). Symposium on Q-fever. Wash. Med. Sci. publ., 6, 28.

[194] Fishbein, D.B., and Raoult, D. (1992). A cluster of Coxiella burnetii infections associated with exposure to vaccinated goats and their unpasteurized products. Am. J. Trop. Med. Hyg., 47, 35–40.

[195] Fournier, P.E., Casalta, J.P., Habib, G., Messana, T., and Raoult, D. (1996). Modification of the diagnostic criteria proposed by the Duke endocarditis service to permit improved diagnosis of Q fever endocarditis. Am. J. Med., 100, 629–633.

[196] Fournier, P.E., Casalta, J.P., Piquet, P., Tournigand, P., Branchereau, A., and Raoult, D. (1998). Coxiella burnetii infection of aneurysms or vascular grafts: report of seven cases and review. Clin. Infect. Dis., 26, 116–121.

[197] Fournier, P.E., Roux, V., and Raoult, D. (1998). Phylogenetic analysis of SFG rickettsiae by study of the outer surface protein rOmpA. Int. J. Syst. Bacteriol., 48, 839–849.

[198] Fournier, P.E., Marrie, J.J., and Raoult, D. (1998). Diagnosis of Q fever. J. Clin. Microbiol., 36 (7), 1823–1834.

[199] Fox, J.G., Anderson, G., Lynn, C., Lower, F.M., and Quimby, F.W., eds. (2002).: Laboratory Animal Medicine 2nd edition, Am. Coll. of Lab. Anim. Med. Ser., Academic press.

[200] Franz, D.R., Jahrling, P.B., Friedlander, A.M., McClain, D.J., and Hoover, D.L. (1997). Clinical recognition and management of patients exposed to biological warfare agents. JAMA, 278, 399–411.

[201] Friedland, J.S., Jeffrey, I., Griffin, G.E., Booker, M., and Cortney-Evans, R. (1994). Q-fever and intrauterine death. Lancet, 343, 288.

[202] Fritz, E. (1994). Inauguaral-Dissertation, Fachbereich Veterinärmedizin, Justus-Liebig Univ., Giessen, Germany, 172 pp.

[203] Fritz, E., Thiele, D., Willems, H., and Wittenbrink, M.M. (1995). Quantification of C. burnetii by PCR and a colometric microtiter plate hybridization assay (CMHA). Eur. J. Epidemiol., 11, 549–557.

[204] Gaon, J.Z. (1958). Gall. Higijena, Belgrad, 10, 200. (по Сербезов, В., Д. Шишманов, Е. Александров, В. Новкиришки. Рикетсиозите в България и другите балкански страни. „Хр. Г. Данов" – Пловдив, 1973).

[205] Gikas, A., Froudarakis, M., Nicolaidis, G., and Tselentis, Y. (1994). Myopericarditis complicating Q fever. Ann. Med. Interna, 145, 149–151.

[206] Gikas, A., Spyridaki, I., Scoulica, E., Psaroulaki, A., and Tselentis, Y. (2001). In vitro susceptibility of C. burnetii to Linezolid in comparision with its susceptibilities to Quinolones, Doxycycline and Clarithromycin. Antimicrob. Agents and Chemotherapy, 45 (11), Nov. 2001, 3276–3278.

[207] Gil-Grande, R., Aguado, J.M., Pastor, C., M. Garcia-Bravo, C. Gomes-Pellico, Soriano, F., and Noriega, A. (1995). Conventional viral cultures and shell vial assay for diagnosis of apparently culture-negative C. burnetii endocarditis. Eur. J. Clin. Microbiol. Infect. Dis., 14, 64–67.

[208] Gimenez, D.F. (1964). Staining rickettsiae in yolk sac cultures. Stain technol., 30, 135–137.

[209] Gofiin, Y.A., van Hoeck, R., and Jachan, E. (2000). Banking of cryopreserved heart valves in Europe assesment of a 10 years. J. Heart Valve Dis., 2, 207–214.

[210] Gonder, J.C., Kishimoto, R.A., Kostello, M.D., Pedersen, C.E., and Larson, E.W. (1979). Cynomolgus monkey model for experimental Q fever infection. J. infect. Dis., 139, 191–196.

[211] Gračea, E., Voiculescu, R., and Zarnea, G. Ztschr. (1970). Immunitätsforsch. Allergie Klin. Immunol., 140, 358 (по Gračea, E. et al. Immunization of a man with a soluble Q fever vaccine. Arch. Roum. de Pathol. Exp. Microbiol, 32, (1973). 45–51).

[212] Graham, C.J., Yamauchi, T., and Rountree, P. (1989). Q-fever in animal laboratory workers: an outbreak and its investigation. Am. J. Inf. Con., 17, 345–348.

[213] Gray, G.C., Rodier, G.R., and Matras-Maslin. V.C. (1995). Serological evidence of respiratory and rickettsial infections among Somali refugees. Am. J. Trop. Med. Hyg., 52 (4), 349–353.

[214] Greenslade, E., Beasley, R., Jennings, L., Woodwarde, A., and Weinstein, P. (2003). Has Coxiella burnetii been introduced into New Zealand? Emerg. Infect. Dis. Jan. 2003, 1–4.

[215] Greiff, D., Prowege, E., and Pinkerton, N. (1957). J. Exp. Med., 105, 105.

[216] Greiner, T.C., Mitros, F.A., Sapleton, J. and van Rybroek, J. (1992). Fine-needle aspiration findings of the liver in a case of Q fever. Diagn. Cytopathol., 8, 181–184.

[217] Guigno, D., Coupland, B., Spith, E.G., Farell, I.D., Desselberger, U., and Caul, E.O. (1992). Primary humoral antibody response to Coxiella burnetii, the causative agent of Q fever. J. Clin. Microbiol., 30, 1958–1967.

[218] Hackstadt, T., and Williams, J.C. (1981). Biochemical stratagem for obligate parasitism of eukaryotic cells by Coxiella burnetii. Proc. Natl. Acad. Sci., USA, 78, 3240–3244.

[219] Hackstadt, T., and Williams, J.C. (1983). Ph dependence of the C. burnetii glutamate transport system. J. Bacteriol., 154, 598–603.

[220] Hacktadt, T., Peacock, M.G., Hitchcock, P.J., and Cole. R.l. (1985). Lipopolysaccharide variation in C. burnetii: intrastrain heterogenecity in structure and antigenicity. Infect. Immun., 48, 359–365.

[221] Hackstadt, T. (1986). Antigenic variation in the phase I lipopolisaccharide of C. burnetii. Infect. Immun., 52, 337–340.

[222] Hackstadt, T. (1988). Steric hindrance of antibody binding to surface proteins of C. burnetii by phase I lipopolysaccharide. Infect. Immun., 56, 802–807.

[223] Hackstadt, T. (1990). The role of lipopolysaccharide. Ann. of the New York Academy of Sci., 590, 27–32.

[224] Hamedeh, G.N., Turner, B.W., and Trible, W. (1992). Laboratory outbreak of Q fever. J. Fam. Pract., 35, 683–685.

[225] Hamzic, S., Beslagić, E., and Zvidić, S. (2003). Significance of Q fever serologic diagnosis of clinically suspect patients. Ann. of the New York Acad. Sci., 990, 365–368.

[226] Harris County Rabies/Animal Control. Q-fever, (2005). Houston, Texas. http://www.cdc.gov./ncidod/EID/vol1no4-1036.htm

[227] Harvey-Sutton, P.L. (1995). Post Q fever syndrome. Med. J. Aust., 162, 168.

[228] Hatchette, T.F., Hudson, R.C., Schlech, W.F., Campbell, N.A., Hatchette, J.F., Ratnam, S., Raoult, D., Donovan, C., and Marrie, T.J. (2001).

Goat-associated Q fever: a new disease in Newfoundland. CDC, Emerging Infect. Dis., 7 (3), 15p., http://www.cdc.gov/ncidod/eid/vol7no3/hachette/htm

[229] Hazard prevention and control in the work environment: airborne dust. World Health Organization, 1999.

[230] Health Safety Executive Statistics, Table RIDDOR01: Cases of occupational disease reported under RIDDOR (a). Sheffield, UK: Health & Safety Executive, 2002.

[231] Health Protection Agency, UK. Q-fever, 2005. http://www.hpa.org.uk/infections/topics/az/zoonoses/qfever

[232] Health Protection Agency, UK. Q-fever, Apr. 2005. http://www.hpa.org.uk/inf/topics/az/zoonoses/qfever/geninfo.htm

[233] Heinzen, R.A., and Mallavia, P. (1987). Cloning and functional expression of the C. burnetii citrate synthase gene of E. coli. Infect. Immun., 55, 847–855.

[234] Heinzen, R.A., Stiegler, G.L., Whiting, L., Schmitt, S.A., Malavia, L.P., and Frazier, M.E. (1990). Use of pulsed field gel electrophoresis to differentiate C. burnetii strains. N.Y. Acad. Sci., 590, 504–513.

[235] Heinzen, R.A., Fraizer, M.E., and Malavia, L.P. (1992). Coxiella burnetii superoxide dismutase gene: clonning, sequencing, and expression in E. coli. Infect. Immun, 60, 3814–3823.

[236] Heinzen, R.A., Scidmore, M.A., Rockey, D.D., and Hankstadt, T. (1996). Differential interaction with endocytic and exocytic pathways distinguish parasitphorous vacuoles of Coxiella burnetii and Chlamydia trachomatis. Infect. Immun., 64, 796–809.

[237] Hendrix, L., and Malavia, L.P. (1984). Active transport of proline by C. burnetii. J. Gen. Microbiol., 130, 2857–2863.

[238] Hendrix, L.R., Samuel, J.E., and Malavia, L.P. (1990). Identification and cloning of a 27-kDa C. burnetii immunoreactive protein. Ann. N.Y. Acad. Sci., 590, 534–540.

[239] Hendrix, L.R., Samuel, J.E., and Malavia, L.P. (1991). Differentiation of C. burnetii isolates by analysis of restriction-endomuclease-digested DNA separated by SDS-PAGE. Gen. Microbiol., 137, 269–276.

[240] Heron, C. (1992). The networks of botanical creation. New scientist, 133, (1807), 32–36.

[241] Hellebrand, W., Breuer, T., and Petersem, L. (2001). Changing epidemiology of Q fever in Germany, 1997–1999. Emerging Infect. Dis., 7 (5), Sept–Oct 2001, 1–15.

[242] Higgins, D., and Marrie, T.J. (1991). Seroepidemiology of Q fever among cats in New Brunswick and Prince Edward Island. Ann. N.Y. Acad. Sci., 590, 271–274.

[243] Hilbink, F., Penrose, M., Kovacova, E., and Kazar, J. (1993). Q fever is absent from New Zealand. Int. J. Epidemiol., 22, 945–949.

[244] Ho, T., Khin Khin Htwe, Yamasaki, N., Zhang Guo Quan, Ogawa, M., Yamaguchi, T., Fukushi, H., and Hirai, K. (1995). Isolation of Coxiella burnetii from dairy cattle and ticks, and some characteristics of the isolates in Japan. Microbiol. Immunol., 39 (9), 663–671.

[245] Hofmann, C.E., and Heaton, J.W. (1982). Q fever hepatitis: clinical manifestation and pathological findings. Gastroenterology, 83, 474–479.

[246] Honstettre, A., Ghigo, E., Moynault, A., Capo, C., Toman, R., Akira, S., Takenchi, O., Lipidi, H., Raoult, D., and Mege, J.L. (2004). Lipopolysaccharide from C. burnetii is involved in bacterial phagocytosis, filamentous actin reorganization, and inflammatory responses through toll-like receptor 4. J. Immunol., 172, 3695–3703.

[247] Hoover, T.A., and Williams, J.C. (1990). Chracterization of Coxiella burnetii pyrB. Ann. N.Y. Acad. Sci., 590, 485–490.

[248] Hoover, T.A., Vodkin, M.H., and Williams, J. (1992). A Coxiella burnetii repeated DNA element resembling a bacterial insertion sequence. J. Bacteriol., 174, 5540–5548.

[249] Hore, D. (1970). Austral. Vet. J., 46 (4), 169–172. (по Генчев, Г., Д. Огнянов, Н. Ганчев. Епизоотологично и епидемиологично проучване на Ку-треската в крупен овцевъден комплекс. Сб. III Нац. конг. Микробиол., II част, (1975). 243–248).

[250] Hotta, A., Zhang Guo Quan, Andoh, M., Yamaguchi, T., Fukushi, H., and Hirai, K. (2004). Use of monoclonal antibodies for analyses of Coxiella burnetii major proteins. J. Vet. Med. Sci., 66 (10), 1289–1291.

[251] Houwer, D.J., Meer van der, M., Djik van, A.A.H.M. (1992). Prevalence and incidence of C. burnetii infections in dogs and cats in the Netherlands and the region of Midden-Holland. Utrecht: Rijksuniversiteit Utrecht facultei Diergeneeskundcl.

[252] Htwe, K.K., Amano, K., Sugiyama, Y., Yagami, K., Minamoto, M., Hashimoto, A., Yamaguchi, T., Fukushi, H., and Hirai, K. (1992). Seroepidemiology of C. burnetii in domestic and companion animals in Japan. Vet. Rec., 131 (21), 490.

[253] Hwang, Y.M., Lee, M.C., Suh, D.C., and Lee, W.Y. (1993). Coxiella (Q fever) – associated myelopathy. Neurology, 43, 338–342.

[254] IACUC Health & Safety in the Vivarium. Q-fever, (2005). http://www.research.uncc.edu/comp/qfever.cfm

[255] Ignatovich, V.F., Lukin, E.P., Makhlai, A.A., and Perepelkin, V.S. (1999). The laboratory diagnosis of rickettsiosis. Voenn. Med. Zh., 9, 5–9.

[256] Igwa, M., de Hertogh, G., Neuville, B., Roskams, T., Nevens, F., and van Steenbergen, W. (2001). Hepatic fibrin-ring granulomas in granulomatous hepatitis: report of four cases and review of the literature. Acta Clin. Belg., 56 (6), Nov–Dec, 341–348.

[257] Illinois Dept. of Public Health. Emergency preparedness. Q-fever, (2005). http://www.jdph.state.il.us/bioterrorism/factssheets/qfever.htm

[258] Imhäuser, K. (1948). Klin. Wschr., 26, 337. (по Imbauser, K. Viruspneumonien: Q-fiebre and virusgrippe, Klin. Wschr., 27, (1949). 353–362).

[259] Imhäuser, K. (1949). Viruspneumonien: Q-fiebre and virusgrippe. Klin. Wschr., 27, 353–362.

[260] Izzo, A.A., and Marmion, B.P. (1993). Variations in interferon-gamma responses to C. burnetii antigens with lympocytes from vaccinated or naturally infected subject. Clin. Exp. Immunol., 94, 507–515.

[261] Janbon, F., Raoult, D., Reynes, J., and Bertrand, A. (1989). Concominant human infection due to Rickettsia conori and Coxiella burnetii. J. Infect. Dis., 160, 354–355.

[262] Janigan, D.T., and Marrie, T.J. (1983). An inflammatory pseudotumor of the lung in Q fever pneumonia. N. Engl. J. Med., 308, 86–87.

[263] Jaspers, U., Thiele, D., and Krauss, H. (1994). Monoclonal antibody based competitive ELISA for the detection of specific antibodies against Coxiella burnetii from different animals species. Zentralblatt für bacteriologie, 281 (1), 61–66.

[264] Jay-Russell, M., Douglas, J., Drenzek, C., Stone, J., Blythe, D., Weltman, A., Jones, T., Craig, A., McQuinston, J., Paddock, C., Nicholson, W., Thompson, H.A., Zaki, S., Wright, J., O'Reilly, M., and Kirschke, D. (2002). Q fever—California, Georgia, Pensylvania and Tennessee, 2000–2001. CDC, MMWR, 51 (41), 924–927.

[265] Julvez, J., Michault, A., and Kerdelhue, C. (1997). Serological study of rickettsia infections in Niamey, Niger. Med. Trop. (Mars), 57 (2), 153–156.

[266] Kaltenboeck, B., Kousoulas, K.G., and Storz, J. (1992). Two-step polymerase chain reactions and restriction endonuclease analyses detect and differentiate ompA DNA of Chlamydia spp. J. Clin. Microbiol., 30 (5), May 1098–1104.

[267] Kang, Y.B., Youn, H.J., and Cho, S.N. (1993). Q fever: detection and correlation of indirect immunofluorescence antibodies in bovine sera to Coxiella burnetii phase I and phase II antigens. RDA, and Agricult, J., Sci., Veterinary, 35 (1), 669–678.

[268] Kang, Y.B., Youn, H.J., and Cho, S.N. (1996). Detection and correlation of indirect immunofluorescence antibodies in bovine sera to Coxiella burnetii phase I and phase II antigens. The Korea-Bulgaria, Int. Seminar on Q-fever and Chlamydiosis, Apr 15–24, Anyang, R. of Korea, 8.

[269] Kaplan, M.M., and Bertagna, P. (1955). The geographical distribution of Q fever. Bull. WHO, 13, 829–860.

[270] Kaulfuss, K. (1997). Infectious abortions in sheep. Man is endangered. InfecDeutsch schafzucht, 89, 406–409.

[271] Kazar, J., Brezina, R., and Schramek, S. (1974). Acta Virol., 18, 91. (byPandarov, S., K. acabalieva. Q-fever, Zemizdat, 1985).

[272] Kazar, J., and Brezina, R. (1991). Control of rickettsial diseases. Eur. J. Epidemiol., 7, 282–286.

[273] Kazar, J. (1999). Q-fever—current concept. In: Rickettsiae and Rickettsial disases at the Turn of the third millenium, Elsevier, 304–319.

[274] Khaschabi, D., and Brandstätter, A. (1994). Serological survey for antibodies to C. burnetii and C. psittaci in sheep in Tyrol, Austria. Wiener Tieräztl. Mon., 81 (10), 290–294.

[275] Khaschabi, D., Schöpf, K., and Schmid, E. (1996). Serological prevalence of Q fever in western Austria. Wien. Tieräztl. Mon., 83 (1), 2–5.

[276] Khavkin, T., and Tabibzadeh, S.S. (1988). Histologic immunofluorescence and electron microscopic study of infectious process in mouse lung after intranasal challenge with C. burnetii. Infect. Immun., 56, 1792–1799.

[277] Kim, S.G, Kim, E.H., Lafferty, C.J., and Dubovi, E. Coxiella burnetii in bulk tank milk samples, United states. CDC, Emerging Infect. Dis., 11 (4), 2005 Apr, http://www.cdc.gov/ncidod/EID/vol1no4-1036.htm

[278] Komiya, T., Sadamasu, K., Kang, M., Tsuboshima, S., Eukushi, H., and Hirai, K. (2003). Seroprevalence of Coxiella burnetii infections among cats in different living environments. J. Vet. Med. Sci., 65 (9), 1047–1048.

[279] Kölb, J. (1992). Comperative study of the diagnosis and prophylaxis of Chlamydia and Coxiella infections of farm animals in Namibia. Inaugural-Dissertation, Fachbereich Veterinärmedizin, Justus-Liebig-Universität, Giessen, Germany, 85.

[280] Kordova, N. (1960). Study of antigenicity and immunogenicity of filterable particles of Coxiella burnetii. Acta Virol., 4, 56–62.

[281] Kordova, N., and Kovačova, E. (1968). Acta Virol., 12, 23.

[282] Kortepeter, M.G., Christopher, G., and Cieslak, T. (2001). Med. Management of Biol. Casnalties Handbook. US Army Med. Res. Inst. Inf. Dis., US Dept. of Defense, 33–36.

[283] Krauss, H. (1989). Clinical aspects and prevention of Q fever in animals. Eur. J. Epidemiol., 5, 454–455.

[284] Krt, B. (2003). The influence of Coxiella burnetii phase I and phase II antigens on the serological diagnosis of Q fever in cattle. Slov. Vet. Res., 40 (3/4), 203–207.

[285] Krumbiegel, E.R., and Wisniewski, H.J. (1970). Q fever in Milwaukee. II. Consumption of infected raw milk by human volunteers. Arch. Environ. Health, 21, 63–65.

[286] La Scola, B., Lepidi, H., and Raoult, D. (1997). Pathological changes during acute Q fever: influence of the route of infection and inoculum size in infected guinea pigs. Infect. Immun., 65, 2443–2447.

[287] La Scola, B., Lepidi, H., Maurin, M., and Raoult, D. (1998). A guinea pig model for Q fever endocarditis. J. Infect. Dis., 178, 278–281.

[288] Lang, G.H. (1990). Coxiellosis (Q-fever) in animals. In: Q-fever, vol. 1. The disease (Marrie, T.J., ed.). CRC press, Inc., Roca Raton, Fla.

[289] Langley, J.M., Marrie, T.J., Covert, A., Waag, D.M., and Williams, J.C. (1988). Poker player's pneumonia: an urban outbreak of Q fever following exposure to a parturient cat. N. Engl. J. Med., 178, 319, 354–356.

[290] Laughlin, T., Wang, D., Williams, J., and Marrie, T.J. (1991). Q fever: from deer to dog to man. Lancet, 337, 676–677.

[291] Lennette, E.H., Holmes, M.A., and Abinanti, F.R. (1961). Am. J. Hyg., 73 (1), 105. (по Огнянов, Д., Н. Павлов, М. Панова, Г. Генчев. Експерим ентална Ку-рикетсиоза по агнетата. Вет. мед. науки, XIII, 8, (1976) 3–9).

[292] Lennox, J. (1993). Ass. Prof. of Med. Emory University School of med., Act. Med. Director, Infect. Dis. Program, Grady Med. Hospital, Atlanta, GA. Personal communication, May 1993. (по Byrne, W.R. In: Med. aspects of Chemical and Biological Warfare. US Dept. of the Army, (1997). 523–537).

[293] Lev, B., Shachaar, A., Segev, S., Weiss, P., and Rubinstein, E. (1988). Quiescent Q fever endocarditis exacerbated by cardiac surgery and corticosteroid therapy. Arch. Intern. Med., 148, 1531–1532.

[294] Levente, S. (2004). Abortion caused by Coxiella burnetii in a sheep. Magyar Allatorvosok Lapja, 126 (8), 482–486.

[295] Leyk, W., and Krauss, H. (1974). Demonstration of submicroscopic reproduceable units of Coxiella burnetii. DTW, 81, 528–530.

[296] Liebisch, A. (1981). Control of D. marginatus in Germany. In: Whitehead, G.B., Gibson, J.D., eds.) Proc. Int. Conf. on Tick Biology and Control under the Auspices of the Tick Res. Unit. Rhodes Univ. S. Africa, Crahamstown, 157–158.

[297] Lim, W.S., Macfarlane, J.T., and Colthorpe, C.L. (2001). Respiratory diseases in pregnancy (bullet) 2: Pneumonia and pregnancy. Thorax, 56, 398–405.

[298] Lupoglazoff, J.M., Brouqui, P., Magnier, S., Hvass, U., and Casoprana, A. (1997). Q fever tricuspid valve endocarditis. Arch. Dis. Child., 77, 5448–5449.

[299] Lipton, J.H., Fong, T.C., Gill, M.J., Burgess, K., and Elliott, P.D. (1987). Q fever inflammatory pseudotumor of the lung. Chest, 92, 756–757.

[300] Literak, J. (1994). Coxiella burnetii antibodies in calves concentrated in a large-capacity calf house in an area with endemic incidence of Q fever. Acta Vet. (Czech Republic), 63 (2), 65–69.

[301] Literak, J. (1995). Acta Vet. (Czech Republic), 40 (3), 77–80. (по Literak, I., J. Řeháček. Q fever—distribution and importance of this disease in the Chech Republic and in the Slovak Republic. Vet. Med., 41 (2), (1996). 45–63).

[302] Literak, J., and Řeháček, J. (1996). Q fever—distribution and importance of this disease in the Chech Republic and in the Slovak Republic. Vet. Med., 41 (2), 45–63.

[303] Literak, J., and Kroupa, L. (1998). Herd-level Coxiella burnetii seroprevalence was not associated with herd-level breeding performance in Chech dairy herds. Preventive Vet. Med., 33 (1/4), 261–265.

[304] Ludlam, H., Wreghitt, T.G., Thornton, S., Thomson, B.J., Bishop, N.J., Coombler, S., and Cunniffe, J. (1997). Q fever in pregnancy. J. Infect., 34, 75–78.

[305] Lyytikäinen, O., Zieze, T., B. Schwartländer, Matzdorff, B., Kuhnhen, C., Burger, C., Krug, W., and Petersen, L. (1997). Outbreak of Q fever in Lohra-Rollshausen, Germany, Spring, 1996. Eurosurveillance, 2 (2), 9–11.

[306] Lyytikäinen, O., Zieze, T., Schwartländer, B., Matzdorff, B., Kuhnhen, C., and Jäger, C. (1998). An outbreak of sheep—associated Q fever

in a rural community in Germany. Eurosurveillance J. Epidemiol., 14, 193–199.

[307] Macchiavello, A. (1937). Rev. Chilean hyg. Med., 1, 101. (по Macchiavello, A. Eine neue Rickettsienfarbung, Zbl. Bakt. Ref., 139, (1941). 291).

[308] Madariaga, M.G., Pulvirenti, J., Sekosan, M., Padock, C.D., and Zaki, S.R. (2004). Q fever endocarditis in HIV-infected patient. Emerg. Infect. Dis. (serial online), 10 (3), March, www.cdc.gov/ncidod(EID)/101/03971.htm

[309] Madariaga, M.G. (2004). CDC—Emerg. Infect. Dis. J. 10 (3), 1–8.

[310] Magill, A. (1994). Medical corps, US Army Infect. Dis. Officer, Dept. Immunol. Div. of Communicable Dis. & Immunol. Walter Reed Inst. of Res., W. Reed Army Med. Center, Washington, DC. Personal communication, Sept.

[311] Mallavia, L.P., Whiting, L.L., Minnick, M.F., Heinzen, R.A., Reschke, D., Foreman, M., Baca, O.G., and Frazer, M.E. (1990). Strategy for detection and differention of Coxiella burnetii strains using polymerase chain reaction. Ann. NY Acad. Sci., 590, 572–581.

[312] Mallavia, L.P. (1991). Genetic of rickettsiae. Eur. J. Epidemiol., 7, 213–221.

[313] Maltezou, H.C., Constantopoulou, I., Kallergi, C., Vlahou, V., Georgakopoulos, D., Kafetzis, D.A., and Raoult, D. (2004). Q fever in children in Greece. Ann. J. Trop. Med. Hyg., 70, 540–544.

[314] Mantovani, A., and Benazzi, P. (1953). The isolation of Coxiella burnetii from Rhipicephalus sanguineus on naturally infected dogs. J. Am. Vet. Med. Assoc., 122, 117–120.

[315] Manual Mult. Species. Q-fever, (2005). http://www.spc.int/rahs/manual/multiplespecies/Qfever.htm

[316] Manuel Cruz, J., Martinez, R., Varela, A., and Hernandez, R. (1994). Pericarditis and Q fever. Ann. Med. Interna, 11, 515.

[317] Marmion, B.P., Shannon, M., Maddocks, I., Storm, P., and Pentilla, I. (1996). Protracted fatique and debility after aqute Q fever. Lancet, 347, 978–979.

[318] Marrie, T.J., Schlech, W.F., Williams, J.C., and Yates, L. (1986). Q fever pneumonia associated with exposure to wild rabbits. Lancet. I., 427–429.

[319] Marrie, T.J., Durant, H., Williams, J.C., Mintz, E., and Waag, D.M. (1988). Exposure to parturient cats is a risk factor for acquisition of Q fever in Maritime Canada. J. Infect. Dis., 158, 101–108.

[320] Marrie, T.J. (1988). Liver involvement in acute Q fever. Chest., 94, 896–898.

[321] Marrie, T.J., Langille, D., Papukna, V., and Yates, Y. (1989). An outbreak of Q fever in a truck repair plant probably due to aerosols from clothing contraminated by contract with a newborn kitten. Epidem. Inf., 102, 119–127.

[322] Marrie, T.J., Stein, A., Janigan, D., and Raoult, D. (1996). Route of infection determines the clinical manifestations of acute Q fever. J. Infect. Dis., 173, 484–487.

[323] Marrie, T.J. (1990). Epidemiology of Q-fever, In: Q-fever (Marrie, T.J., ed.,), The disease, CRC Press, Bota Racon, Fla, vol. 1, 49–70.

[324] Marrie, T.J. (1990). Coxiella burnetii (Q-fever). In: Principles and practice of infect. Dis., (Mandell, G.L., J.E. Bennett eds.), 3rd ed Churchill Livingstone, Inc., New York, 1727–1734.

[325] Marrie, T.J. (1993). Q fever in pregnancy: report of two cases. Infect. Dis. Clin. Pract., 2, 207–209.

[326] Marrie, T.J. (1995). Coxiella burnetii (Q fever) pneumonia, Clin. Infect. Dis., 21 (Suppl. 3). 5253–5264.

[327] Marrie, T.J., and Pollak, T.P. (1995). Seroepidemiology of Q fever in Nova Scotia: evidence for age dependent cohorts and geographical distribution. Eur. J. Epidemiol., 11, 47–54.

[328] Marrie, T.J., and Raoult, D. (1997). Q fever—a review and issues for the next century. Int. J. Antimicrob. Agents, 8, 145–161.

[329] Marrie, T.J. (2003). Coxiella burnetii pneumonia. Eur. Respire. J., 21, 713–719.

[330] Marrie, T.J. (2004). Q fever pneumonia. Curr. Opin. Infect. Dis., 17 (2), Apr., 137–142.

[331] Martin, J., and Innes, P. (2002). Q-fever. Ontario Ministry of agriculture and food. Sept., http://www.gov.on.ca/OMAFRA/english/livestock/vet/facts/info-qfever. htm

[332] Martini, M., Baldelli, R., and Paulucci de Calboli, L. (1994). Ztbl. für Bakteriologie, 280 (3), 416–422.

[333] Martinov, S., Schoilev, C., and Popov, G. (1989). Chlamydien abort bei schafen in der Volkrepublik Bulgarien. Mh. Vet. Med. (Berlin), 44 (11), 361–400.

[334] Martinov, S.P., and Popov, G.V. (1992). Electronmicroscopic diagnosis of Chlamydial infections. Int. Symposium of WAVMI, Davis, Ca., USA, 8–12 Sept. 1992, 394.

[335] Martinov, S.P., Tcherneva, E., and Jersek, B. (1996). Repetitive extra-genic palindromic sequence polymorphism based polymerase chain reaction (rep-PCR) is a useful tool in diagnostics of C. psittaci infection in sheep and studies of field and vaccinal chlamydial strains. Infect. Dis. In Obstetrics and Gynecology, 4 (3), 180–183.

[336] Masala, G., Porcu, R., Sanna, G., Chessa, G., Cillara, G., Chisu, V., and Tola, S. (2004). Occurrence, distribution and role of abortion of C. burnetii in sheep and goats in Sardinia. Vet. Microbiol., 99 (3/4), 301–305.

[337] Masuzawa, T., Sawaki, K., Nagaoka, H., Akiyama, M., and Hirat, K.Y (1997). Yanagihara. Relationship between pathogenicity of C. burnetii isolates and gene sequence of the macrophage infectivity potentiator (Cbmip) and sensor-like protein (qrsA). FEMS Microbiol. Letters, 154 (2), 201–205.

[338] Matthewman, L.A., and Rycroft, A.N. (1996). Exposure of cats in southern Africa to Coxiella burnetii, the agent of Q fever. South Africa Med. J., 86 (1), 94–95.

[339] Matthewman, L.A., Kelly, P., Hayter, D., Downie, S., Wray, K., Bryson, N., Rycroft, A., and Raoult, D. (1997). Exposure of cats in southern Africa to C. burnetii, the agent of Q fever. Eur. J. Epidemiol., 13 (4), 477–479.

[340] Maurin, M., Benoliet, A.M., Bongrand, P., and Raoult, D. (1992). Phagolysosomes of Coxiella burnetii—infected cells maintain an acid pH during persistent infection. Infect. Immun., 60, 5013–5016.

[341] Maurin, M., Benoliet, A.M., Bongrand, P., and Raoult, D. (1992). Phagolysosomal alkanization and the bactericidal effect of antibiotics: the Coxiella burnetii paradigm. J. Infect. Dis., 166, 1097–1102.

[342] Maurin, M., and Raoult, D. (1997). Bacteriostatic and bactericidal activity of levofloxacin against Rickettsia rickettsii, Rickettsia conori, "Israeli spotted group rickettsia" and Coxiella burnetii. J. Antimicrob. Chemother., 39, 725–730.

[343] Maurin, M., and Raoult, D. (1999). Q fever. Clin. Microbiol. Rev., 12 (4), 518–533.

[344] McCaul, T.F., Hackstadt, T., and Williams, J.C. (1981). Ultrastructural and biological aspects of C. burnetii under physical disruptions. In: Rickettsiae and rickettsial diseases (W. Burgdorfer, Anacker, R.L., eds.). Academic press, Inc., New York, 267–280.

[345] McCaul, T.F., and Williams, J.C. (1981). Developmental cycle of Coxiella burnetii: structure and morphogenesis of vegetative and sporogenic differentiations. J. Bacteriol., 147, 1063–1076.

[346] McCaul, T.F. (1991). The development cycle of C. burnetii, In: Q-fever: the biology of C. burnetii (Williams, J.C., Thompson, H.A., eds.), CRC Press. Inc. Boca Raton, Fla., 223–258.

[347] McCaul, T.F., Williams, J.C., and Thompson, H.A. (1991). Electron microscopy of Coxiella burnetii in tissue culture. Induction of cell types as products of developmental cycle. Acta Virol., 35, 545–556.

[348] McCaul, T.F., Dare, A.J., Gannon, J.P., and Galbraith, A.J. (1994). In vivo endogenous spore formation by Coxiella burnetii in Q fever endocarditis. J. Clin. Pathol., 47, 978–981.

[349] McQuiston, J.H., Childs, J.E., and Thompson, H.A. (2002). Q fever. J. Am. Vet. Med. Assoc., 221, 796–799.

[350] McQuiston, J.H., Gibbons, R.V., Velic, R., Nicholson, W.L., Castrodale, L., Wainright, S.H., Vanniewenhoven, T.J., Morgan, E.W., Arapovic, L., Delilic, A., O'Reilly, M., and Bajrovic, T. (2003). Investigation of focus of Q fever in a nonfarming population in the Federation of Bosnia and Herzegovina. Annals NYAS online, 990, 229–232.

[351] McQuiston, J.H., Nargund, V.N., Miller, J.D., Priestley, R., Shaw, E.I., and Thompson, H.A. (2003). Prevalence of antibodies to Coxiella burnetii among Veterinaty School Dairy Herds in the United states, Vector-Borne and Zoonotic Dis., Mar. 2005, 5 (1), 90–91.

[352] Medical monitoring program for Q-fever. EH & S Biosafety Office, 7.10.2002, http://www.ehs.ufl.edu/bio/qfever/qmedsurv.htm

[353] Medline plus medical encyclopedia: Q-fever, (2004). http://www.nlm.nih.gov/medline-plus/ency/articll/001337.htm

[354] Mege, J.L., Maurin, M., Capo, C., and Raoult, D. (1997). Coxiella burnetii: the query fever bacterium. A model of immune subversion by a strictly intracellular microorganism. FEMS Microbiol. Rev., 19, 209–217.

[355] Migala, A.F. (2004). Q-fever (eMedicine), 1–15, http://www.emedicine.com/ped/topic1973.htm

[356] Mihaljevič, F. (1949). Zbornik radova I kongresa lekara FNRY, Beograd, Med. kniga, 3, 269.

[357] Miller, M. (2005). McGill University—Q-fever (Coxiella burnetii) protocol, rev. Oct. 2005, 1–8.

[358] Millonig, G. (1961). Vorteila eines Phosphatpuffers für OsO_2— Lösungen bei der Fixierung. J. Appl. Phys., 32, 1637.

[359] Mo, Y.Y., and Mallavia, L.P. (1994). A Coxiella burnetii gene encodes a sensor-like protein. Gene, 15, 185–190.

[360] Moll, van, P., Baumgärtner, W., Eskens, U., and Hänichen, T. (1993). Immunocytochemical demonstration of Coxiella burnetii antigen in the fetal placenta of naturally infected sheep and cattle. J. Comp. Pathol., 109 (3), 295–301.

[361] Moore, J.D., Barr, B.C., Daft, B.M., and O'Conor, M.T. (1991). Pathology and diagnosis of Coxiella burnetii infection in a goat herd. Vet. Pathol. 28 (1), 81–84.

[362] Morita, C., Katsugama, J., Yanase, T., Veno, H., Muramatsu, Y., Hohdatsu, T., and Koyama, H. (1994). Seroepidemiological survey of Coxiella burnetii in domestic cats in Japan. Microbiol. Immunol., 38, 1001–1003.

[363] Moos, A., Vishwanath, S., and Hackstadt, T. (1988). Experimental Q-fever endocarditis in rabbits. In: Abs 7th Nat. Meet. Am. Soc. Rickettsiae & Rickettsial Dis., Santa Fe, M, N.

[364] Mosquera Losano, J.D., Brea Hernando, A.J., Lopez Bonilla, A., and Marquez de Prado Urquita, M. (1994). Pleuropericarditis caused by Q fever. Ann. Med. Interna, 11, 366–367.

[365] Moughabghab, A.V., Fenides, A., Woelffle, D., and Socolovsky, C. (1994). Coxiella burnetii pleuropericarditis. Rev. Med. Interne, 15, 838–840.

[366] Moustaghfir, A., Deharo, J.C., Dupont, H.T., Macaluso, G., Raoult, D., and Djiane, P. (1995). Acute pericarditis in Coxiella burnetii infection. Arch. Mal. Coeur. Vaiss., 88, 1657–1659.

[367] Mölle, G., Hentschke, J., and Laiblin, C. (1995). Diagnosis of Q fever in a flock of sheep in Berlin. J. Vet. Med. Ser. B., 42 (7), 405–413.

[368] Murray, E.S., Djakovič, P., Ljupša, F., and Snyder, J.C. (1951). Publ. Health. Rep., 66, 1032.

[369] Musso, D., and Raoult, D. (1995). Coxiella burnetii blood cultures from acute and chronic Q fever patients. J. Clin. Microbiol., 33, 3129–3132.

[370] Musso, D., Drancourt, M., Osscini, S., and Raoult, D. (1996). Sequence of quinolone resistance-determining region of gyrA gene for clinical isolates and for an in vitro-selected quinolone—resistant strain of C. burnetii. Antimicrob. Agents Chemother., 40, 870–873.

[371] Myers, W.F., Baca, O.G., and Wisseman, C.L. (1980). J. Bact., 144 (1), 460–461.

[372] Nagaoka, H., Akiyama, M., Sigieda, M., Nishio, T., Akahane, S., Hattori, H., Ho, T., Fukushi, H., and Hirai, K. (1996). Isolation of

Coxiella burnetii from children with influenza—like symptoms in Japan. Microbiol. Immunol., 40, 147–151.

[373] Nagaoka, H., Sugieda, M., Akiyama, M., Hishina, T., Akahane, S., and Fujiwara, K. (1998). Isolation of Coxiella burnetii from vagina of feline clients at veterinary clinics. J. Vet. Med. Sci., 60 (2), 251–252.

[374] National Q-fever Management Program. Australian Governmenr Dept. of Health and Ageing, Apr 2004, http://immunise.health.gov/au/qfever/index.htm

[375] National Q-fever Management Program. Population Health, Dept of Health, Australia, (2005). http://www.population.health.wa.gov.au/communicable/qfever.cf.m

[376] Neil, M., and Li, W.H. (1985). Mathematical model for studying genetic variation in terms of restriction endonucleases. Proc. Natl. Acad. Sci. USA, 76, 5269–5273.

[377] Nermut, M.V., Scheamek, S., and Bresina, R. (1968). Electron microscopy of Coxiella burnetii phase I and II. Acta Virol., 12, 446–452.

[378] Nochimson. G. CBRNE—Q-fever. eMedicine, (2004). 1–7, http://www.emedicine.com/EMERG/topic492.htm

[379] Nguyen, S.V., Otsuka, H., Zhang, G.O., To, H., Tamaguchi, T., Fukushi, H., Noma, A., and Hirai, K. (1996). Rapid method for detection of Coxiella burnetii antibodies using high-density particle agglutination. J. Clin. Microbiol., 34, 2947–2951.

[380] OIE—Institute for International Cooperation in Animal Biologies—Center for Food Security & Public Health (CFSPH): Query fever, (2004). http://www.cfsph.iastate.edu

[381] OIE Terrestrial Manual. Ch. 2.2.10., Q-fever, (2004). 387–398.

[382] Ontario Ministry of Health and Long Term care. Q-fever, (2002). http://www.health.gov.on.ca/english/public/pub/disease/qfever.html

[383] Ossewaarde, J.M. (1992). PCR for Chlamydia trachomatis. J. Clin. Microbiol., 30, 2122–2128.

[384] Osterman, V.J., and Parr, P.R. (1974). Infect. Immun., 4 (5), 1152–1155.

[385] Paiba, G.A., Green, L.E., Lloyd, G., Patel, D., and Morgan, K.L. (1999). Prevalence of antibodies to Coxiella burnetii (Q fever) in bulk tank milk in England and Wales. Vet. Rec., 144 (19), 519–522.

[386] Palmer, N.C., Kierstaed, M., and Key, D.W. (1983). Placentitis and abortions in goats and sheep in Ontario caused by Coxiella burnetii. Can. Vet. J., 24, 60–63.

[387] Paretsky, D. (1990). The biology of Coxiella burnetii and the patho-biochemistry of Q fever and its endotoxicosis. Ann. of the New York Acad. of Sciences, 5, 417–421.

[388] Pattacini, O., Massiro, I., and Lodetti, E. (1994). Q fever a probable emerging zoonosis: seroepidemiological study in the Reggio Emilia province, Italy. Proc. 18th World Buiatrics Congress, Bologna, Aug 29—Sept 2, 845–848.

[389] Payzin, S., and Golem, B. (1948). The presence of Q fever in Turkey. Trop. Dis. Bull., 45, 10, 891.

[390] Payzin, S. (1950). Epidemiology of Q fever in Turkey. Trop. Dis. Bull., 47, 2, 129.

[391] Peacock, M.G., Philip, R.N., Williams, J.C., and Faulkner, R.S. (1983). Serological evaluation of Q fever in humans: enhanced phase I titers of immunoglobulines G and A are diagnostic for Q fever endocarditis. Infect. Immun., 41, 1089–1098.

[392] Perez Ortola, R., Fernandes Penalba, G., Cia Ruiz, J., and Merino, A. (1992). Osteorthritis por Coxiella burnetii. Rev. Clin. Esp., 191, 25–26.

[393] Perez-del-Molino, A., Aguado, J.M., Riancho, A., Sampredo, I., Mator- ras, P., and Gonzales-Macias. J. (1991). Erythromycin and the treat- ment of Coxiella burnetii pneumonia. J. Antimicrob. Chemother., 28, 455–459.

[394] Perry, S., Dennie, C.J., Koblenz, C.L., and Cleland, S. (1994). Mini- mizing the risk of Q fever in the hospital setting. Can. J. Infect. Control, 9, 5–8.

[395] Peter, O., Dupuis, G., Burgdorfer, W., and Peacock, M. (1985). Evalu- tion of the complement fixation and indirect immunofluorescence tests in the early diagnosis of Q fever. Eur. J. Clin. Microbiol. Infect. Dis., 4, 394–396.

[396] Peter, O., Dupuis, G., Bee, D., Luthy, R., Nicolet, J., and Burgdorfer, W. (1988). Enzyme-linked immunosorbent assay for diagnosis of chronic Q fever. J. Clin. Microbiol., 26, 1978–1982.

[397] Peter, O., Flepp, M., Bestetti, G., Nicolet, J., Luthy, R., and Dupuis, G. (1992). Q fever endocarditis: diagnostic approaches and monitoring of therapeutic effects. Clin. Invest., 70, 932–937.

[398] Philip, C.B. (1948). Comments of the name of the Q-fever organisms. Public Health Rep., 63, 58.

[399] Pinsky, R.L., Fishbein, D.B., Greene, C.R., and Gensheimer, K.F. (1991). An outbreak of cat-associated Q fever in the United States. J. Infect. Dis., 164 (1), 202–204.

[400] Piquet, P., Raoult, D., Tranier, P., and Mercier, C.I. (1994). Coxiella burnetii infection of pseudoaneurism of an aortic bypass graft with contiguous vertebral osteomyelitis. J. Vasc. Surg., 19, 165–168.

[401] Pitcher, D.G., Sanders, N.A., and Owen, R.J. (1989). Rapid extraction of bacterial genomic DNA with Guanidium thiocyanate. Letters Appl. Microbiol., 8, 151–156.

[402] Plavsič, Ž. (1992). Valjanost antigena C. burnetii dobijenih različitum methodania I njihova dijagnostička vrednost u RVK. Disertacija, Beograd.

[403] Plavsič, Ž., Ašanin, R., and Valčić, M. (1993). Q-groznica—immunološke metode. Izd. VSI Kraljevo, s. 74.

[404] Popov, G.V., and Martinov, S.P. (1980). Morphologie und Morphogenese einiger Chlamydienstämme. Wiss. Z. der Humboldt Universitat zu Berlin, Math-Nat. R., XXIX, 1, 11–17.

[405] Preston, P.M., Yin Hong (eds.). (1997). Proc. Europ. Union Inter. Symp. on tick and tick-borne diseases. Trop. Animal Health and Production, 29 (4 suppl), 144.

[406] Puchades, P.S., de Medrano, V.A.L., Gonzalez, E.O., Bosco, J.L.P., and Ballester, A.H. (2005). Q-fever endocarditis, http://sprojects.mmi.mcgi ll.ca/heart/mar990526rl.html

[407] Purdue University. Animal Exposure Program. Ruminant handling and q-fever prevention, (2005). http://www.purdue.edu/rem/eh/amm luse.htm

[408] Q-fever, Child and Youth Health—CYH (South Australia), (2005). http://www.healthinside.gov.au/topics/qfever

[409] Q-fever. In: Manual of standarts for diagnostic tests and vaccines. Office Int des epizooties, Paris, Fourth edition, (2000). 822–831.

[410] Q-fever. In: The Merck Vet. Manual, 8th ed. (Aiello, S.E., Mays, A., eds.), Whitehouse Station, NJ: Merck and Co, (1998). 486–487.

[411] Q-fever, (2004). http://www.arches.uga.edu/~aeneary/qindex.htm

[412] Q-fever guidelines. ULACC, (2005). http://animal.tamu.edu/qfever.htm

[413] Q-fever (Query fever), (2005). http://www.ehs.ufl.edu/bio/qfever/ qinfo.htm

[414] Queensland Government, Dept. of Primary Industries and Fisheris, (2005). Zoonotic diseases—Q-fever, http://www.dpi.qld.gov.au/ ealth/3925.html

[415] Rady, M., Glavits, R., and Nagy, G. (1985). Demonstration in Hungary of Q fever associated with abortions in cattle. Acta Vet. Hung., 33, 169–176.

[416] Raoult, D., Bollini, G., and Gallas, H. (1989). Osteoarticular infection due to Coxiella burnetii. J. Infect. Dis., 159, 1159–1160.

[417] Raoult, D., Drancourt, M., and Vestris, G. (1990). Bactericidal effect of doxycycline associated with lysosomotropic agents on Coxiella burnetii in P388D1 cells. Antimicrob. Agents Chemother., 34, 1512–1514.

[418] Raoult, D., Vestris, G., and Enea, M. (1990). Isolation of 16 strains of C. burnetii from patients by using a sensitive centrifugation cell culture system and establishment of strains in HEL cells. J. Clin. Microbiol., 28, 2482–2484.

[419] Raoult, D. (1990). Host factors in the severity of Q fever. Ann. N.Y. Acad. Sci., 590, 33–38.

[420] Raoult, D., Torres, H., and Drancourt, M. (1991). Shell-vial assay: Evaluation of a new technique for determing antibiotic susceptibility, tested in 13 isolates of Coxiella burnetii. Antimicrob. Agents Chemother., 35, 2070–2077.

[421] Raoult, D., Brouqui, P., Marchon, B., and Gastaut, J.A. (1992). Acute and chronic Q fever in patients with cancer. Clin. Infect. Dis., 14, 127–130.

[422] Raoult, D., Levi, P.Y., Tissot Dupond, H., Chicheportiche, C., Tamalet, C., Gastaut, J.A., and Salducci, J. (1993). Q fever and HIV infection. AIDS, 7, 81–86.

[423] Raoult, D., Stein, A., and Engl, N. (1994). Q fever during pregnancy, a risk for women, fetuses and obstetricians. J. Med., 330, 371.

[424] Raoult, D., and Marrie, T. (1995). Q fever. Clin. Infect. Dis., 20, 489–496.

[425] Raoult, D., Houpikian, P., Tissot Dupond, H., Riss, J.M., Arditi-Djiane, J., and Brouqui, P. (1999). Comparision of 2 regiments containing doxycycline and ofloxacin or hydroxychloroquine. Arch. Intern. Med., 159, 167–173.

[426] Raoult, D., H. Tissot-Dupond, Foucault, C., and Gouvernet, J. (2000). Clinical and epidemiological features of 1383 infections (see comments). Medicine (Baltimore), 2, 109–123.

[427] Raoult, D., Mege, J.L., and Marrie, T. (2001). Q-fever: queries remaining after decades of research. In: Emerging Infections, 5, (Scheld, W.M., Craig, W.A., Hudges, J.M., eds.), Washington DC, AMS Press, 29–56.

[428] Raoult, D., Fenollar, F., and Stein, A. (2002). Q fever during pregnancy. Arch. Intern. Med., 162, 701–704.

[429] Rauch, A.M., Tanner, M., Pacer, R.E., Barrett, M.J., Brokopp, C.D., and Schonberger, L.B. (1987). Sheep-associated outbreak of Q fever— Idaho. Arch. Intern. Med., 147, 341–344.

[430] Řehaček, J., Brezina, R., and Majerska, M. (1968). Multiplication of rickettsiae in tick cells in vitro. Acta. Virol., 12, 41–43.

[431] Řehaček, J., Krauss, H., Kocianova, E., E. Kovačova, Hinterberger, E., Hanak, P., and Tuma, V. (1993). Studies of the prevalence of C. burnetii, the agent of Q fever, in the foothills of the southern Bavarian Forest, Germany. Zentralblatt fur bakteriologie, 278 (1), 132–138.

[432] Reilly, S., Northwood, J.L., and Caul, E.O. (1990). Epidemiol. Infect., 105, 391–408.

[433] Reintjes, R., Hellenbrand, W., and Düsterhaus, A. (2000). Q-Fieber— Ausbruch in Dortmund im Sommer 1999. Gesumdh-Wes., 62, 1–6.

[434] Relman, D.A., Loutit, J.S., Schmidt, T.M., Falkow, S., and Tomkins, L.S. (1990). The agent of bacillary angiomatosis. An approach to the identification of uncultured pathogens. N. England J. Med., 323, 1573–1580.

[435] Rexroth, G., Rasch, W., and Altmannsberger, M. (2000). Bone marrow granulomatosis in Q fever. Med. Klin., 7, 404.

[436] Reynolds, E.S. (1963). Die Verwendung von Bleizitrat bei hohem pH ais eine electronendichte Färbung in der Elecronenmikroscopie. J. Cell. Biol., 17, 208–217.

[437] Rikinisa, Y. (1991). The tribe Erlichieae and erlichial disease. Clin. Microbiol. Rev., 4, 286–308.

[438] Rilley, J.M., Campbell, W.E., and Patrick, W.C. (1964). J. Bacteriol., 88, 802.

[439] Ripoll, M.M., Gomez-Mendierta, M.A., J. Gomes-Cerezo, Belichon, J.C., and Molina, F. (1997). Pericarditis recidicante por fiebre Q. Enferm. Enferm. Infecc. Microbiol. Clin., 15, 168.

[440] Robbins, F.C., Gauld, R.L., and Warner, F.B. (1946). Q fever in the Mediterranean area: Report of its occurrence in Allied troops, II. Epidemiology. Am. J. Hyg., 44, 23–50.

[441] Rohde, C., Kelly, P.J., and Raoult, D. (1993). Dairy cows as reservoirs of Coxiella burnetii in Zimbabwe. Centr. African J. Med., 39 (10), 208–210.

[442] Rolain, J.M., Maurin, M., Vestris, G., and Raoult, D. (1998). Antibiotic susceptibilities of 26 rickettsiae to 13 antibiotics. Antimicrob. Agents Chemother., 42 (7), Jul., 1537–1541.

[443] Rolain, J.M., Maurin, M., and Raoult, D. (2001). Bacteriostatic and Bactericidal activities of Moxifloxacin against C. burnetii. Antimicrob. Agents Chemother., 45, 301–302.

[444] Roman, M.J., Coriz, P.D., and Baca, O.G. (1986). A proposed model to explain persistent infection of host cells with Coxiella burnetii. J. Gen. Microbiol., 132, 1415–1422.

[445] Rose, M., Wemheuer, W., and Schmidt, F.W. (1994). Follow-up of IgG$_1$, antibodies to C. burnetii in blood and milk of cattle with particular reference to the reproductive cycle and milk yield. DTW, 101 (12), 484–486.

[446] Rotz, L., Khan, A., and Lillibridge, S. (2002). Public health assessment of potential biological terrorism agents. Emerg. Infect. Dis., 8, 225–230.

[447] Rudakov, N.V. (1996). Ecologico-epidemiological observation of Q-fever in Russia. In: Rick. & Rickettsial dis. (J. Kazar, Toman, R., eds.), Sl. Acad. Sci., Bratislava, 524–527.

[448] Russo, P. (1997). Infection à C. burnetii ou Fievre Q. In: Manual pratique de diagnostic de laboratoire des avortements infect. des petits ruminants (Rodolakis, A., and Nettleton, P. eds.), L'Espace Veterinaire, Casablanca, 103–114.

[449] Ruiz-Contreras, J., Montero, R.G., and Ramos, J.T. (1993). Q fever in children. AJDC, 147, 300–302.

[450] Rustscheff, S., Norlander, L., Macellaro, A., Sjostedt, A., Vene, S., and Carlsson, M. (2000). A case of Q fever acquired in Sweden and isolation of the probable etiological agent, Coxiella burnetii from an indigenous source. Scand. J. Infect. Dis., 32, 605–607.

[451] Saiki, R.K., Gelfand, D.H., Stoffel, S., Schart, S.J., Higuchi, R., Horn, G.T., Mullis, K.B., and Erlich, H.A. (1988). Primer directed enzymatic amplification of DNA with a thermostable DNA polymerase. Science, 23, 487–491.

[452] Salmon, M., Howells, B., Glencross, E., Evans, A., and Palmer, S. (1982). Q fever in an urban area. Lancet, 1, 1002–1004.

[453] Sambrook, J.E., Fritsch, F., and Maniatis, T. (1989). Molecular cloning: a laboratory manual., 2nd ed., Cold Spring Harbor Lab., Cold Spring Harbor, N.Y.

[454] Samuel, J.E., Frazier, M.E., Kahn, M.L.E., Thomshow, L.S., and Malavia, L.P. (1983). Isolation and characterization of a plasmid from phase I Coxiella burnetii. Infect. Immun., 41, 448–493.

[455] Samuel, J.E., Frazier, M.E., and Malavia, L.P. (1985). Correlation of plasmid type and disease caused by Coxiella burnetii. Infect. Immun., 49, 775–779.

[456] Sanford, S.E., Josephson, G.K.A. and MacDonald, A. (1993). Q fever abortions in a goat herd. Can. Vet. J., 34 (4), 246.

[457] Sanford, S.E., Josephson, G.K.A., and MacDonald. A. (1994). A Cox-iella burnetii (Q fever) abortion storms in goat herds after attendance at an annual fair. Can. Vet. J., 35 (6), 376–378.

[458] Sanchis, R., Russo, P., Calamel, M., and Pepin, M. (1997). Diagnostic differential des avortements infect. des petits ruminants. In: Manual pratique de diagnostic de lab. des avortments infect. des petits rumi-nants (Rodolakis, A., and Nettleton, P. eds,), L'Espace veterinaire, Casablanka, 25–34.

[459] Sanzo, J., M.A. Garcia-Calabuig, Audicana, A., and Dehesa, V. (1993). Q fever: prevalence of antibiotics to Coxiella burnetii in the Basque Country. Int. J. Epidemiol., 22, 1183–1187.

[460] Savinelli, E.A., and Mallavia, L.P. (1990). Comparision of C. bur-netii plasmid to homologous chromosomal sequences present in a plasmidless endocarditis—causing isolate. Ann. N.Y. Acad. Sci., 590, 523–533.

[461] Schmeer, N., P. Müller, Langel, J., Krauss, H., Frost, J.W., and Wieda, J. (1987). Q fever vaccines in animals. Zentbl. Bacteriol. Microbiol. Hyg. Ser. A., 267, 79–88.

[462] Schmeer, N., Wieda, J., Frost, W., Herbst, W., Weiss, R., and Krauss, H. (1987). Erfahrungen bei der diagnose, differentialdiagnose and bekämfung des bovinen Q-fiebers in einem vorzugsmilchbestand mit fruchtbarkeisstörungen. Vet. Immunol. Immunopathol., 15, 311–322.

[463] Schramek, S., and Mayer, H. (1982). Different sugar compositions of lipopolysaccharides isolated from phase I and phase II cells of C. burnetii. Infect. Immun., 38, 53–57.

[464] Schramek, S., Radziejewski-Lebrect, J., and Mayer, H. (1985). 3-C-branched aldoses in lipopolysaccharide of phase I Coxiella burnetii and their role as immunodominant factors. J. Biochem., 455–461.

[465] Schröder, H.D. (1998). Berl. Münch. Tierärztl. Woch., 111 (5), 173–174.

[466] Scott, G.H., and Williams, J.C. (1990). Susceptibility of Coxiella burnetii to chemical disinfectants. Ann. N.Y. Acad. Sci., 590, 291–296.

[467] Scrimgeour, E.M., Johnston, W.J., and Al Dhahry, S.S.H. (2005). First report of Q fever in Oman. Medscape, http://www.medscape.com/viewarticle/414698

[468] Šeguljev, Z., Vidič, B., Šajalik, M., and Groič, Z. (1997). The impor-tance of sheep in the epidemiology of Q fever in Voivodina. Folia Veterinaria, 41 (3/4), 107–111.

[469] Selvaggi, T.M., Rezza, G., Scagnelli, M., Rigoli, M., Rassu, M., F. de Lalla, Pellizzer, G.P., Tramrin, A., Bettini, C., Zampieri, L., Belloni, M.,

Dalla Pozza, E., Marangon, S., Marchioreto, N., Togni, G., Giacobbo, M., Todescato, A., and Bankin, N. (1996). Investigation of Q fever outbreak in northern Italy. Eur. J. Epidemiol. 12 (4), 403–408.

[470] Serbezov, V., Alexandrov, E., Matova, E., Mateva, M., Georgiev, G., Ganchev, N., and Genchev, G. (1978). Some data on ecology of C. burnetii and Spotted fever group rickettsiae in Bulgaria. Rep. 7th Int. Congr. Infect. and Parasit. Dis., Oct, Varna, I, 511–514.

[471] Serbesov, V., Kasar, J., Novkirishki, V., Gatcheva, N., Kovachova, E., and Voynova, V. (1999). Q fever in Bulgaria and Slovakia. Emerg. Infect. Dis., 5 (3), 388–394.

[472] Seshadri, R., Paulsen, I.T., Eisen, J.A., Real, T.D., Nelson, K.E., Nelson, W.C., Ward, N.L., Tettelin, H., Daugherty, S.C., Brinkac, L.M., Maduri, R., Dodson, R.J., Khouri, H.M., Lee, K.H., Carty, H.A., Scanlan, D., Heinzen, R.A., Thompson, H.A., Samuel, J.E., Fraser, C.M., and Heidelberg, J.F. (2003). Complete genome sequence of the Q fever pathogen Coxiella burnetii. Proc. Natl. Acad. Sci. USA, 100 (9), Apr 29, 5455–5460.

[473] Shaw, E.I., Moura, H., Woolfitt, A.R., Ospina, M., Thopson, H.A., and Barr, J.R. (2004). Anal. Chem., 76 (14), 4017–4022.

[474] Sheridan, P., MacCaig, J.N., and Hart, R.J.C. (1974). Myocarditis complicating Q fever. Br. Med. J., 2, 155–156.

[475] Sidwell, R.W., Thorpe, B.D., and Gebhardt, L.P. (1964). Studies of latent Q fever infections. I. Effects of whole body X irradiation upon latently infected quinea pigs, white mice, and deer mice. Am. J. Hyg., 79, 113–124.

[476] Sidwell, R.W., Thorpe, B.D., and Gebhardt, L.P. (1964). Studies of latent Q fever infections. II. Effects of multiple cortisone injections. Am. J. Hyg., 79, 320–327.

[477] Siegman-Igra, Y., Kaufman, O., Keysary, A., Rzotkiewicz, S., and Shalit, I. (1997). Q fever endocarditis in Israel and a worldwide review. Scand. J. Infect. Dis., 29, 41–49.

[478] Smith, D.L., Ayres, J.G., Blair, I., Burge, P.S., Carpenter, M.J., Caul, E.O., Coupland, B., Desselberger, U., Evans, M., and Farell, I.D. (1993). A large Q fever outbreak in the West Midlands: clinical aspects. Respir. Med., 87, 509–516.

[479] Sobradillo, V., Zalacain, R., Capelastegui, A., Uresandi, F., and Corral, J. (1992). Antibiotic treatment in pneumonia due to Q fever. Thorax., 47, 276–278.

[480] Socal, R.R., and Sneath, P.H.A. (1963). In: Principles of Numerical Taxonomy. Ed. Freeman, San Francisco, Ca, 169–210.

[481] Soliman, A.K., Botros, B.A.M., and Watts, D.M. (1992). Evaluation of a competitive enzyme immunoassay for detection of Coxiella burnetii antibody in animal sera. J. Clin. Microbiol., 30 (6), 1595–1597.

[482] Somma-Moreira, R.E., Caffarena, R.M., Somma, S., G. Pèrez, and Monteiro, M. (1987). Analysis of Q fever in Uruguay. Rev. Infect. Dis., 9, 386–387.

[483] Spicer, A.J. (1978). Military significance of Q-fever. A review J.R. Soc. of Med., 71, 762–767.

[484] Spicer, A.J., Peacock, M.G., and Williams, J.C. (1981). Effectiveness of several antibiotics in suppressing chick embryo lethality during exp. infect. by C. burnetii, R. typhi. and R. rickettsii. In: Rickettsiae and ricketts. dis. (W. Burgdorfer, Anacker, L., eds.), Acad. Press. Inc., New York, 375–383.

[485] Spelman, D.W. (1982). Q fever: a study of 111 consecutive cases. Med. J. Aust., 1, 547–553.

[486] Spyridaki, I., Gikas, A., Kofteridis, D., Psaroulaki, A., and Tselentis, Y. (1998). Q fever in the Greek island of Crete: detection, isolation, and molecular identification of eight strains of C. burnetii from clinical samples. J. Clin. Microbiol., 36 (7), 2063–2067.

[487] Stamp, J.T., Watt, J.A., and Cockburn, R.B. (1952). Enzootic abortion in ewes. J. Comp. Pathol. Ther., 62 (2), 93–101.

[488] Stein, A., and Raoult, D. (1992). A simple method for L amplification of DNA from paraffin-embedded tissues. Nucleic Acids Res., 20, 5237–5238.

[489] Stein, A., and Raoult, D. (1992). Detection of Coxiella burnetii by DNA amplification using polymerase chain reaction. J. Clin. Microbiol., 30 (9), 2462–2466.

[490] Stein, A., Saunders, N.A., Taylor, A.G., and Raoult, D. (1993). Phylogenic homogenicity of Coxiella burnetii strains as determined by 16S ribosomal RNA sequencing. FEMS Microbiol. Lett., 113, 339–344.

[491] Stein, A., and Raoult, D. (1993). Lack of pathotype specific gene in human Coxiella burnetii isolates. Microbiol. Pathog., 15, 177–185.

[492] Stein, A., and Raoult, D. (1995). Q fever endocarditis. Eur. Heart. J., 16 (Suppl. B), 19–23.

[493] Stein, A., Kruszewska, D., Gouvernet, J., and Raoult, D. (1997). Eur. J. Epidemiol., 13 (4), 471–475.

[494] Stein, A., and Raoult, D. (1998). Q fever during pregnancy: a public health problem in Southern France. Clin. Infect. Dis., 27, 592–596.

[495] Stein, A., and Raoult, D. (1999). Pigeon pneumonia in province: a bird-borne Q-fever outbreak. Clin. Infect. Dis., 29 (3), Sep, 617–620.

[496] Stein, A., Louveau, C., Lepiai, H., Ricci, F., Baylac, P., Davoust, B., and Raoult, D. (2005). Q fever pneumonia: virulence of Coxiella burnetii pathovars in a murine infection. Infect. Immun., Apr, 73 (4), 2469–2477.

[497] Sting, R., and Westphal, K. (1993). Purification of Coxiella burnetii antigen by gelchromatography for use in enzyme-linked immunosorbent assay (ELISA). J. Vet. Med. Series B., 40 (9/10), 690–696.

[498] Sting, R., Kopp, J., Mandl, J., Seeh, C., and Seeman, G. (2003). Comparative serological studies in dairy herds of Chlamydia and Coxiella burnetii infections using CFT and ELISA. Tier. Umschau, 58 (10), 518–528.

[499] Sting, R., Breitling, N., Oehme, R., and Kimmig, P. (2004). Studies on the prevalence of Coxiella burnetii in sheep and ticks of the genus Dermacentor in Baden-Wuer-Hemberg. DTW, 111 (10), 390–394.

[500] Stoenner, H.G., and Lackman, D.B. (1960). The biological properties of Coxiella burnetii isolated from rodents collected in Utah. Am. J. Hyg., 71, 45–48.

[501] Stoker, M., and Fiset, P. (1956). Phase variation of the Nine Mile and other strains of R. burnetii. Can. J. Microbiol., 2, 310–321.

[502] Storz, J. (1971). Chlamydia and Chalmydia—induced diseases. C.C. Thomas. Publ. House, Springfield, Ill., USA, 358 pp.

[503] Srigley, J.R., Vellend, H., Palmer, N., Phillips, M.J., Geddie, W.R., von Nostrand, A.W., and Edwards, V.D. (1985). Am. J. Surg. Pathol., 9, 752–758.

[504] Suhan, M., Chen, S.Y., Thompson, H.A., Hoover, T.A., Hill, A., and Williams, J.C. (1996). Cloning and characterization of an autonomous replication sequence from Coxiella burnetii to ampicillin resistance. J. Bacteriol., 178, 5233–5243.

[505] Suputtamongkol, Y. (2003). Q fever in Thailand (Letter to the editor). Emerg. Infect. Dis., 9 (1), 1–4.

[506] Tcherneva, E., Rijpens, N., Naydensky, C., and Herman, L.M.F. (1996). Repetitive element sequence based polymerase chain reaction for typing of Brucella strains. Vet. Microbiol., 51, 169–178.

[507] Tcherneva, E.K., Ljutzkanov, M., and Ivanov, P. (1999). Typing of Brucella species by application of different PCR techniques. Bulg. J. Vet. Med., 2 (1), 17–32.

[508] Tcherneva, E., Rijpens, N., Jersek, B., and Herman, L.M. (1999). Differentiation of Brucella species by randomly amplified polymorphic DNA analysis. J. Appl. Microbiol.

[509] Texas Department of Health Zoonosis Control Division, Q-fever, (2004). 6, http://www.tdh.state.tx.us/zoonosis/diseases/qfever

[510] The Merck Manual of Diagnosis and Therapy. (2005). Q-fever, sec. 13, chap. 159.

[511] Thiele, D., Willems, H., G. Köpf, and Krauss, H. (1993). Polymorphism in DNA restriction patterns of C. burnetii isolates investigated by PAGE and image analysis. Eur. J. Epidemiol., 9, 419–425.

[512] Thiele, D., Willems, H., Haas, M., and Krauss, H. (1994). Analysis of the entire nucleotide sequence of the cryptic plasmid QpH$_1$ from Coxiella burnetii. Eur. J. Epidemiol., 10, 413–420.

[513] Thiele, D., and Willems, H. (1994). Is plasmid based differentiation of Coxiella burnetii in "acute" and "chronic" isolates still valid? Eur. J. Epidemiol., 10, 427–434.

[514] Thomas, D.R., Salmon, R.L., Smith, R.M.M., Caul, E.O., Treweek, L., Kench, S.M., Coleman, T.J., Meadows, D., P. Morgan-Carner, and Sillis, M. (1996). Epidemiology of Q-fever in the UK. In: Rickettsiae and rickettsial diseases (J. Kazar, Toman, R., eds.). Slovac Acad. Sci., Bratislava, 512–517.

[515] Thoms, H.J. (1996). Epidemiologishe Untersuchungen zum Vorkommen von C. burnetii auf vier Truppenübungsplätzen der Bundeswehr in Nordrhein-Westfalen und Niedersachsen (diss.), Giessen, Germany, Justus-Liebig—Universität.

[516] Tigertt, W.D., and Benenson, A.S. (1956). Studies on Q fever in man. Trans. Assoc. Am. Phys., 69, 98–104.

[517] Tissot Dupont, H., Raoult, D., Brouqui, P., Janbon, F., Peyramond, D., Weiller, P.J., Chicheportiche, C., Nezri, M., and Poirier, R. (1992). Epidemiological features and clinical presentation of acute Q fever in hospitalized patients—323 French cases. Am. J. Med., 93, 427–434.

[518] TissotDupont, H., Thirion, X., and Raoult, D. (1994). Q fever serology: cutoff determination for microimmunofluorescence. Clin. Diagn. Lab. Immunol., 1, 189–196.

[519] Tissot Dupont, H., Torres, E., Nezri, M., and Raoult, D. (1999). A hyperendemic focus of Q fever related to sheep and wind. Amer. J. Epidemiol., 150, 67–74.

[520] To, H., Htwe, K.K., and Hirai, K. (1995). Prevalence of C. burnetii antibodies in dairy cows with reproductive disorders in Japan. Dept. of Vet. Microbiol., Faculty of agriculture, Gifu University, Gifu, Japan.

[521] To, H., Kako, N., Zhang, G.Q., Otsuka, H., Ogawa, M., Ochiai, O., SA V. Nguyen, Yamaguchi, T., Fukushi, H., Nagaoka, N., Akiyama, M., Amano, K., and Hirai, K. (1996). Q fever pneumonia in children in Japan. J. Clin. Microbiol., 34, 647–651.

[522] To, H., Sakai, R., Shirota, K., Kano, C., Abe, S., Sugimoto, T., Takehara, K., Morita, C., Takashima, I., Maruyama, T., Yamaguchi, T., Fukushi, H., and Hirai, K. (1998). Serological investigations and isolation of C. burnetii from wild and domestic birds. J. of Wildlife Dis., 34 (2), 310–316.

[523] Tokarevich, N.K., Schramek, S., and Daiter, A.B. (1990). Indirect haemolysis test in Q fever. Acta Virol., 34, 358–360.

[524] Tselentis, Y., Gikas, A., Kofteridis, D., Kyriakakis, E., Lydatakis, N., Bouros, D., and Tsaparas, N. (1995). Q fever in the Greek island of Grete: epidemiological, clinical, and therapeutic data from 98 cases. Clin. Infect. Dis., 20, 1311–1316.

[525] US Davis, Environmental Health & Safety. Occupational Health and Animals. Q-fever, (2005). http://ehs.ucdavis.edu/animal/health/qfever.cfm

[526] Uhaa, I.J., Fishbein, D.B., Olson, J.G., Rives, C.C., Waag, D.M., and Williams, J.C. (1994). Evaluation of specificity of indirect enzyme-linked immunosorbent assay for diagnosis of human Q fever. J. Clin. Microbiol., 32, 1560–1565.

[527] University of Tennessee. Health Science Center. Q-fever, (2003). http://www.utmem.edu/univheal/OC-Q-fever.z.html

[528] University of Florida, Q-fever Control Policy, Med. Monitoring Program for Q-fever, Oct 2005, http://www.ehs.ufl.edu.bio/qfever/qmain.htm

[529] Urrutia, A., et al. (1991). Acute myopericarditis as a manifestation of Q fever. Med. Clin., 96, 319.

[530] Vachon, P. (1996). Q fever at the animal facilities of a hospital research center, Mèdecine Vèt. du Quebec, 26 (1), 20–22.

[531] Van der Lugt, J., van der Lugt, B., and Lane, E. (2002). Congress of the Mpumalanga branch of the SAVA. Large Animal Section, 1, http://vetpath.vetspecialists.co.za/large1.htm

[532] Von Dopfer, G., Schmeer, N., Frost, J.W., Lohrbach, W., and Wachendörfer, G. (1986). Vetraglichkeits-und immunisierungsversuche

mit einer kommerziellen vakzine gegen Chlamydia psittaci und Coxiella burnetii. DTW, 93, 241–280.

[533] Van Woerden, H.C., Mason, B.W., Nehaul, L.K., Smith, R., Salmon, R., Healy, B., Valappil, M., Westmoreland, D., Martin, S., Evans, M.R., Lioyd, G., Kirkwood, M.H., and Williams, N.S. (2004). Q fever outbreak in industrial setting. Emerg. Infect. Dis. (serial on the Internet), 10 (7), Jul., http://www.cdc.gov/ncidod/EID/vol0no7/03-053.htm

[534] Valkova, D., and Kazar, J. (1995). A new plasmid (QpDV) common to Coxiella burnetii isolates associated with acute and chronic Q fever. FEMS Microbiol. Lett., 125, 275–280.

[535] Versalovich, J., Koeuth, T., and Lupski, J.R. (1991). Distribution of repetitive DNA sequences in Eubacteria and application to fingerprinting of bacterial genomes. Nucleic Acid. Res., 19, 6823–6831.

[536] Vilet, A.H.M. van, Jongejan, F., and Zeijst, B.A.M. (1992). Phylogenetic position of Cowdria ruminantium (Rickettsiales) determined by analysis of amplified 16S ribosomal DNA sequences. Int. J. Syst. Bacteriol., 42, 494–498.

[537] Vodkin, M.H., Williams, J.C., and Stephenson, E.H. (1986). Genetic heterogenicity among isolates of Coxiella burnetii. J. Gen. Microbiol., 132, 455.

[538] Vodkin, M.H., and Williams, J.C. (1986). Overlapping deletion in two spontaneous phase variants of Coxiella burnetii. J. Gen. Microbiol., 132, 2587–2594.

[539] Vodkin, M.H., and Williams, J.C. (1988). A heat shock operon in Coxiella burnetii produces a major antigen homologous to a protein to both mycobacteria and Escherichia coli. J. Bacteriol., 170, 1227–1234.

[540] Von Gouverneur, K., Schmeer, N., and Krauss, H. (1984). Zur epidemiologie des Q-Fiebers in Hessen: Untersunchungen mit dem enzymimmuntests (ELISA) und der komplementbindungsreaction (KBR). Berl. Münch. Tierärztl. Wschr., 97, 437.

[541] Waag, D. (1994). US Army Med. Res. Inst. of Infect. Dis., Pathogenesis and Immunol. Branch, Bacteriol. Div., Fort Detrick, Frederick, Md. Personal communication, Sept.

[542] Waag, D., Chulay, J., Marrie, T., England, M., and Williams, J.C. (1995). Validation of an enzyme immunoassay for serodiagnosis of Q fever. Eur. J. Clin. Microbiol. Infect. Dis., 14, 421–427.

[543] Wachter, R.F., Briggs, G.P., Gangemi, J.D., and Pedersen, C.E. (1975). Changes in buoyant density relationships of 2 cell types of Coxiella burnetii phase I. Infect. Immun., 12, 433–436.

[544] Walker, D.H. (1996). Rickettsiae. In: Med. Microbiology, 4th ed. (S. Baron ed.), York, N., Ch. Livingstone.

[545] Walters, S. (2004). ADDL, Spring, Newsletter, http://www.addle.purdue.edu/newsletters/2004/spring/qfever.htm

[546] Webster, J.P., Lioyd, G., and Macdonald, D.W. (1995). Q-fever (Coxiella burnetii) reservoir in wild brown rat (Rattus norvegicus) populations in the UK. Parasitology, 110, 31–35.

[547] Webster, J.P. (1996). Wild brown rats (Rattus norvegicus) as a soonotic risk on farms in England and Wales. Commun. Dis. Rep., CDR Rev., 6, R 46–49.

[548] Weiss, E., and Moulder, J.W. (1984). Genus III. Coxiella. In: Krieg, N.R., G. Holt (ed.), Bergey's manual of syst. bacteriology, vol. 1 The Willaims & Wilkins Co., Baltimore, Md.

[549] Weisburg, W.G., Woese, C.R., Dobson, M.E., and Weiss, E. (1985). A common origin of rickettsiae and certain plant pathogens. Science, 230, 556–558.

[550] Wesburg, W.G., Dobson, M.E., Samuel, J.E., Dasch, G.A., Malavia, L.P., Baca, O., Mandelco, L., Sechrest, J.E., Weiss, E., and Woese, C.R. (1989). Phylogenic diversity of the Rickettsiae. J. Bacteriol., 171, 4202–4206.

[551] Welsh, J., and McClelland, M. (1990). Fingerprinting genomes using PCR with arbitrary primers. Nucleic Acid Res., 18, 7213–7218.

[552] Wiebe, M.E., Burton, P.R., and Shankel, D.M. (1972). Isolation and characterization of two cell types of Coxiella burnetii phase I. J. Bacteriol., 110, 368–377.

[553] Willems, H., Thiele, D., and Krauss, H. (1993). Plasmid based differentiation and detection of Coxiella burnetii in clinical samples. Eur. J. Epidemiol., 9 (4), 411–418.

[554] Willems, H., Thiele, D., R. Frölich-Ritter, and Krauss, H. (1994). Detection of Coxiella burnetii in cows milk using polymerase chain reaction (PCR). J. Vet. Med. Ser. B., 41 (9), 580–587.

[555] Willems, H., Thiele, D., Burger, C., Ritter, M., Oswald, W., and Krauss, H. (1996). Molecular biology of C. burnetii. In: Rickettsiae and rickettsial diseases (J. Kazar, Toman, R., ed.), Slovak, Acad. of Sci., Bratislava, 363–378.

[556] Willems, H., C. Jäger, and Baljer, G. (1998). Physical and genetic map of the obligate intracellular bacterium C. burnetii. J. Bacteriol., 180, 3816–3822.

[557] Williams, J.C. (1991). In: Q-fever: The biology of C. burnetii (Williams, J.C., Thompson, H.A., eds.), Boca Raton, Fla, CRC Press, chap. 2., 25.

[558] Wilson, K.H., Blitchington, R., Shah, P., McDonald, G., Gilmore, R.D., and Mallavia, L.P. (1985). Probe directed at a segment of Rickettsia rickettsii rRNA amplified with polymerase chain reaction. J. Clin. Microbiol., 27, 2692–2696.

[559] Winn, I.F., Abinanti, F.R., Lennette, E.H., and Welsh, H.H. (1961). Am. J. Hyg., 73 (1), 105 (by Pandarov, S., and Akabalieva, K. Q fever, Zemizdat, 1985).

[560] Winstead, E. (2003). Potential bioweapon: Q fever genome is sequenced, Genome News Network, http://www.genomenewsnetwork. org/articles/05.03/qfever.shtm

[561] Wittenbrink, M., M. Gefäller, Failing, S., and Bisping, K. (1994). Influence of herd and animal factors on detection of complement fixation antibodies against Coxiella burnetii in cattle. Berl. Münch. Tierärz. Wschr., 107 (6), 185–191.

[562] Woernle, H., Limouzin, C., K. Müller, and Durand, M.P. (1985). La fièvre Q bovine. Effects de la vaccination et de làntibiothèrapie sur l'èvolution clinique et l'excrétion de Coxiella dans le lait et les secretions utèrines. Bull. Acad. Vet. de France, 58, 91–100.

[563] Woernle, H., and Müller, K. (1986). Q-fieber beim Rind: Vorkommen, Bekämpfung mit Hilfe der Impflung und/older antibiotischen Behandlung. Tierärz. Umschau, 41, 201–212.

[564] Woldehiwet, Z. (2004). Q-fever (coxiellosis): epidemiology and pathogenesis. Res. in Vet. Sci., 77 (2), 93–100.

[565] Wood dust: hazards and precautions. Woodworking Sheet No. 1 (revised). Sheffield, UK, Helth and Safety Executive, 2001.

[566] Woods, C.R., and Versalovic, J. (1993). J. Clin. Microbiol., 31, 1927–1931.

[567] World Health Organization. Health Aspects of Chemical and Biological Weapons. Reports of a WHO Group of Consultants. Geneva, WHO, 1970.

[568] Yanase, T., Muramatsu, Y., Ueno, H., and Morita, C. (1997). Seasonal variations in the presence of antibodies against Coxiella burnetii in dairy cattle in Hokkaido, Japan. Microbiol. Immunol., 41, 73–75.

[569] Yanase, T., Muramatsi, Y., Inouye, I., Okabayashi, T., Ueno, H., and Morita, C. (1998). Microbiol. Immunol., 42 (1), 51–53.

[570] Yarrow, A., Slater, P.E., and Costin, C. (1990). Q fever in Israel. Publ. Health Rev., 18, 129–137.

[571] Yeaman, M.R., and Baca, O.G. (1990). Unexpected antibiotic susceptibility of a chronic isolate of Coxiella burnetii. Ann. N.Y. Acad. Sci.,
590, 297–305.

[572] Yebra, M., Marazuela, M., Albarran, F., and Moreno, A. (1988). Rev.
Infect. Dis., 10, 1229–1230.

[573] Yoshie, K., Oda, H., Nagano, N., and Matayoshi, S. (1991). Serological
evidence that the Q fever agent (Coxiella burnetii) has spread widely
among dairy cattle of Japan. Microbiol. and Immunol., 35 (7), 577–581.

[574] Yu, X., and Raoult, D. (1994). Serotyping Coxiella burnetii isolates from acute and chronic Q fever patients by using monoclonal
antibodies. FEMS Microbiol. Lett., 117, 15–19.

[575] Yunker, C.E., Ormsbee, R.A., Cory, J., and Peacock, M.G. (1970). Acta
Virol., 14, 383.

[576] Yurtalan, S. (2003). Seroprevalence of Coxiella burnetii (Q fever) infection in cattle in Marmara region. Pendic Vet. Mikrobiyoloji Dergisi, 34
(1/2), 41–49.

[577] Zarnea, G., Vasilik, V., Voiculescu, R., Israel, H., Perederi, S.,
Tunari, C., Szegli, L., Popescu, F., and Ionescu, H. (1958). Trop.
Dis. Bull., 55, 7, 748. (по Сербезов, В., Д. Шишманов, Е. Александров,
В. Новкиришки. Рикетсиозите в България и другите балкански страни.
„Хр. Г. Данов" —Пловдив, 1973).

[578] Zarnea, G., Alexandresco, N., Angelesco, G., Voiculesco, R., Ionesco,
H., and Krausz, N. (1961). Arch. Roum. Path. Exp., 20, 1.

[579] Zhang, G., To, H., Russell, K.E., Hendrix, L.R., Yamaguchi, T., Fukushi,
H., Hirai, K., and Samuel, J.E. (2005). Identification and characterization of an immunodominant 28-kilodalton C. burnetii outer membrane
protein specific isolates associated with acute disease. Infect. Immun.,
73, 1561–1567.

[580] Zhang Guo Quan, Ho To, Yamaguchi, T., Fukushi, H., and Hirai, K.
(1997). Microbiol. and Immunol., 47 (11), 871–877. (by Zhang, G.,
To, H., and Russell, K.E. et al. Identification and characterization of
an immunodominant 28-kilodalton C. burnetii outer membrane protein
specific to isolates associated with acute disease. Infect. Immun., 73,
(2005). 1561–1567).

[581] Zeman, D., Kirkbride, C., Leslie-Steen, C.A., and Duimstra, J.R.
(1989). Ovine abortion due to Coxiella burnetii infection. J. Vet. Diagn.
Invest., 1 (2), 178–180.

[582] Zuber, M., Hoover, T.A., and Court, D.L. (1995). Cloning, sequencing
and expression of the dnaJ gene of Coxiella burnetii. Gene, 152, 99–102.

[583] Zuber, M., Hoover, T.A., and Court, D.L. (1995). Analysis of a Coxiella burnetii gene products that activates capsule synthesis in Escherichia coli: requirements for the heat shock chaperone DnaK and the two-component regulator Rcs. C. J. Bacteriol., 177, 4238–4244.

[584] Zuerner, R.L., and Thompson, H.A. (1983). Protein synthesis by intact Coxiella burnetii cells. J. Bacteriol., 156, 186–191.

[585] Zuzman, T., Yerushalmi, G., and Segal, G. (2003). Functional similarities between the icm/dot pathogenesis systems of Coxiella burnetii and Legionella pneumorphila. Infect. Immun., 71, 3714–3723.

[586] Martinov, S., and Popov, G. (1985). New data on the problem of Q fever in animals. Proc. Scient. Reports of VI Congress of Microbiology, Varna, Oct., Vol. I, 594–601.

[587] Pandarov, S., and Martinov, S. (1981). Serological studies on Q fever in cattle. Vet. Sciences, XVIII, 3, 33–38.

[588] Martinov, S. (1987). Rickettsial zoonoses in Bulgaria. Workshop on surveillance, prevention and control of rickettsial zoonoses in the Mediterranean. MZCP, WHO, Palermo, Italy, June, RICK (87.2), WP7, Ag. Item 3.

[589] Martinov, S.P., Pandarov, S., and Popov, G.V. (1989). Seroepizootology of Q fever in Bulgaria during the last five years. Eur. J. Epidemiol., 5 (4), 425–427.

[590] Martinov, S.P., Neikov, P., and Popov, G.V. (1989). Experimental Q fever in sheep. Eur. J. Epidemiol., 5 (4), 428–427.

[591] Martinov, S. (2006). Serological and virological studies of Q fever in cattle., Vet. Medicine, X, 1–2, 7–14.

[592] Martinov, S., M. Halacheva, Nedelchev, N., Alexandrov, E., and Александров, Е. (2006). Contemporary state of the tick-borne transmissive infections of animals in Bulgaria. Vet. Medicine, X, 3–4, 7–18.

[593] Martinov, S. (2006). Contemporary state of the natural focal animal reservoir of Q fever in Bulgaria. Proc. Scientific. Reports of seventh Nat. Congress of Medical Geography with intl. participation, Sofia, 27–29 Oct. 2005, Vol. I, 323–328.

[594] Martinov, S. (2007). Studies on the Q rickettsial infection in goats. Vet. Medicine XI, 1–2, 18–25.

[595] Martinov, S. (2005). Serological studies and isolation of Coxiella burnetii in sheep. Vet. Medicine, IX, 3–4, 16–22.

[596] Martinov, S. (2007). Animal Q fever in Bulgaria pp. 280–299. In: Contemporary state of the rickettsioses in the world and in Bulgaria

/E. Alexandrov (Ed. In Chief), Kazar, J., Hechemy, K., Kantardjiev, T., Eds/. Sofia, Prof. Marin Drinov Academic Publ. House, 400 pages.

[597] Martnov, S.P. (2006). Studies on some biological, morphological and immunological properties of Coxiella burnetii, the status and peculiarities of natural and agricultural foci of Q fever in Bulgaria. D. Sci. Dissertation, NDRVMI – Sofia, 522 pp.

[598] Martinov, S.P. (2007). Contemporary state of the problem Q fever in Bulgaria. Biotechnol. & Biotechnol. Eq., 21 (3), 353–361.

[599] Teoharova, M., Alexandrov, E., Martinov, S., Kamarinchev, B., Alexandrova, D., Dimitrov, D., Lazarov, H., and Girov, K. (2002). Q fever in Bulgaria – state and problems. Infectology 3, 24–28.

[600] Martinov, S. E. (2006). Alexandrov. Study on Coxiellosis in domestic animals during the last Q fever epidemics in humans in Bulgaria. 12th ISID congress, Lisbon, Portugal, (abs).

[601] Martinov, S. (2008). Coxiellosis in respiratory diseases in sheep. Vet. Medicine, XII, 1–2, 27–36.

[602] Martinov, S. (2007). Studies on mastitis in sheep, caused by Coxiella burnetii. Biotechnol. & Biotechnol. Eq., 21 (4), 484–490.

[603] Van der Hoek, W. (2012). The 2007–2010 Q fever epidemic in the Netherlands: risk factors and risk groups (PhD) Thesis, Utrecht University, ISBN: 978-90-6464-565-5.

[604] Georgiev, M., Afonso, A., Heubawrer, H., et al. (2013). Q fever in humans and farm animals in four European countries, 1982 to 2010. Eurosurveillance, 18 (8): pii= 20407.

[605] Roest H.I., Ruuls, R.S., Tilburg, J.J. et al. (2011). Molecular epidemiology of Coxiella burnetii from ruminants in Q fever outbreak, the Netherlands. Emerg. Infect. Dis., 17 (4), 668–675.

[606] Martinov, S. (2004). Section Three: Veterinary medical geography. In "Medical geography of Bulgaria" – Part I (J. Naumov, ed.), Publisher: Union of Scientists in Bulgaria, Sofia, 213–229.

[607] Martinov, S., and Alexandrov, E. (2005). Serological study of Q rickettsiosis among domestic animals during the epidemic of Q fever in humans in the region of Botevgrad. Vet. Medicine, IX, 1–2, 12–17.

[608] Martinov, S., Teoharova., M., Nedelchev, N., Alexandrova, D., Alexandrov, E., and Kamarinchev, B. (2004). Features in dynamics and the spread of Q fever in Bulgaria. Proc. Scientific reports of Medical Geography, Sofia, 64–68.

[609] Teoharova, M., Martinov, S., and Alexandrov, E. (2002). Serological studies in patients with atypical pneumonia during the epidemic of Q fever in the Etropole region (III–V 2002). Proc. reports of the tenth Congress of the Bulgarian microbiologists with intl. participation, Oct. 10–12, Vol. I, 108–110.

[610] Martinov, S., and Pandarov, S. (2005). Serological study of experimental and natural Q fever in cattle and sheep. Biotechnol. & Biotechnol. Eq., 19 (1), 165–169.

[611] Handjieva, D., Maksimov, K., and Martinov, S. (1985). Q rickettsial infection and epilepsy. Abstr. XII Scientific Session of Primary Clinical Hospital "Prof. Dr. S. Kirkovich" RNDM, Society of Neurology – St. Zagora, Sliven, Yambol 7–8 June 1985, Stara Zagora, 42.

[612] Martinov, S., Bonovska, M., Alexandrov, E., and Tcherneva, E. (2006). Detection and identification of rickettsiae by polymerase chain reaction. Proc. Scientific Reports of Seventh Nat. Congress of Medical Geography with intl. participation, Sofia, 27–29. Oct. 2005, Vol. I, 336–347.

[613] Martinov, S., and Bonovska, M., and Tcherneva, E. (2008). PCR for demonstration and differentiation of coxiellas and rickettsiae. Vet. Medicine, XII, 3–4, 14–20.

[614] Kounchev, M.A. (2009). Development and improvement of PCR for diagnostics of Coxiella burnetii. PhD Dissertation Military Medical Academy – Sofia, 154 pp.

[615] Martinov, S. (2008). Coxiella burnetii Endometritis and Mastitis in Cows. International Journal of Infectious Diseases, 12, Suppl. 1, December 2008, e126.

[616] Pandarov, S., Martinov, S., Petrova, N., and Minkov, V. (1997). Clinical and epizootiological studies on Q-rickettsial endrometrites and mastites in cows andsheep. Compt. Rend. Acad. Bulg. Sci., 50, 5, 141–143.

[617] Vaidya, V.M., Malik, S.V., Bhiegaonkar, K.N., Rathore, R., Kaur, S., and Barbuddhe, S.B. (2010). Prevalence of Q fever in domestic anomals with reproductive disorders. Comp. Immunol. Microbiol. Infect. Dis. 33 (4), 307–321.

[618] Agerholm, J.S. (2013). Coxiella burnetii associated reproductive disorders in domestic animals—aq critical review. Acta Vet. Scandinavica, 55:13.

[619] Khalili, M., Sakhaee, E., and Babaei, H. (2012). Frequency of anti-Coxiella burnetii antibodies in cattle with reproductive disorders. Comp. Clin. Pathol. 21 (5), 917.

[620] Martinov, S. (2007). Agricultural foci and epidemiological characteristics of Q fever. Vet. Medicine XI 3–4, 13–21.

[621] Neykov, P., Martinov, S., and Genchev, D. (1990). Pathomorphological studies in pregnant sheep experimentally infected with R. burnetii, Vet. Sbirka 3, 37–40.

[622] Neykov, P., Martinov, S., and Genchev, D. (1990). Pathomorphological studies in experimentally infected with Rickettsia burnetii rabbits. Vet. Sbirka 2, 20–23.

[623] Van der Brom, P.V. (2009). Q fever outbreaks in small ruminants and people in the Netherlands. Small Ruminant Res., 86 (1–3), 74–79.

[624] Roest, H.I., Tilburg, J.J., Van der Hoek W. et al. (2011). The Q fever epidemic in the Netherlands: history, onset, response and reflexion. Epidemiol. Infect. 139 (1), 1–12.

[625] Martinov, S. (2010). Risk Assessment of Coxiella burnetii in some animal products. Report on International joint course of EFSA – Bulgarian Focal Centre on risk assessment in the food chain, Sept., Sofia.

[626] Martinov, S.P. (1990). Q fever. In: "Viral infections in intensive stock-breeding" (edited by H. a Haralambiev and M. Dilovski), Zemizdat, Sofia, 132–137.

[627] Martinov, S., Alexandrov, E., and Nedelchev, N. (2006). Etiologic and nozogeographic studies of the tick-natural focal reservoir of Q fever in Bulgaria. Proc. Scientific. Reports VII Nat. Congress of Medical Geography with intl. participation, Sofia, 27–28 Oct. 2005, vol. I, 329–336.

[628] Alexandrov, E., Nedelchev, N., Alexandrova, D., Martinov, S., Dimitrova, Z., and Kamarinchev, B. (2005). Natural and agricultural foci of the Marseille and Q-fevers in Bulgaria. 4th Int. Conf. on Rickettsial and Rickettsial Diseases, June 18–21, Longrono (La RIOJA), Spain.

[629] Martinov, S.P., and Pandarov, S. (1999). Investigation of the susceptibility of C. burnetii strains to antibiotics depending of their virulence. II. Susceptibility of C. burnetii as determined by its phase conditions. Biotechnol. & Biotechnol. Eq., 13 (2), 45–50.

[630] Martinov, S.P., and Pandarov, S. (1999). Investigation of the susceptibility of C. burnetii strains to antibiotics depending of their

virulence. I. Isolation of C. burnetii strains and study of their properties. Biotechnol. & Biotechnol. Eq., 13 (1), 73–76.

[631] Martinov, S. (2012). Q fever in animals 277–291. In: "Zoonoses in humans and animals" (S. Martinov and R. Komitova, Eds.), "Medicine and Sports Publishing house", 530 pp.

[632] Martinov, S.P. (2009). Aetiological and epidemiological studies of Q fever and Mediterranean spotted fever in animals. Clinical Microbiology and Infection, 15/4, Suppl. 1:S17, April 2009.

Index

About the Author

Professor Svetoslav P. Marrtinov, DVM, Ph.D., D.Sci. Director of the Central Research Veterinary Medical Institute – Sofia, Bulgaria (1990–2001). Longtime Head of the National Reference Laboratory of Chlamydia and Rickettsia at the institute. He was Head of the Department of especially dangerous infections and zoonoses and Head of the Department of virology and viral diseases at the National Diagnostic and Research Veterinary Medical Institute – Sofia. He works in the fields of microbiology, virology, infectious diseases in animals, and zoonoses. Publications: more than 300 in these directions. Manages or participates in numerous national and international research projects. Over the years, the activities of Professor Martinov also include his work as first Vice-President of the Union of Veterinarians in Bulgaria, founder and first chairman of the Veterinary Association of the countries from Balkan and Black Sea region, member of the Supreme Veterinary Council at the Ministry of Agriculture in Bulgaria, and Vice-President, Bulgarian Society of Medical Geography. He is a member of several national and international scientific societies. Biographical data about Dr. Martinov are included in several editions of the American Biographical Institute, USA, and the International Biographical Centre, Cambridge, England.